THE BARNACLE GOOSE

THE BARNACLE GOOSE

JEFFREY M. BLACK, JOUKE PROP
AND KJELL LARSSON

T & AD POYSER
London

Published 2014 by T & AD Poyser,
an imprint of Bloomsbury Publishing Plc, 50 Bedford Square, London WC1B 3DP.
Reprinted 2015

ISBN (print) 978-1-4729-1157-5
ISBN (epub) 978-1-4729-1156-8

www.bloomsbury.com
www.bloomsburywildlife.com

Bloomsbury is a trademark of Bloomsbury Publishing Plc

Bloomsbury Publishing, London, New Delhi, New York and Sydney

A CIP catalogue record for this book is available from the British Library

Commissioning Editor: Jim Martin

Design by Julie Dando at Fluke Art
Graphics by Dick Visser

Illustrations by Mark Hulme

Printed and bound in India by Replika Press Pvt. Ltd.

10 9 8 7 6 5 4 3 2

Visit bloomsburywildlife.com to find out more about our authors and their books

Contents

We dedicate this book to Rudi Drent and Myrfyn Owen,
for their inspirational role in the study of wild geese,
the Barnacle Goose in particular.

List of Figures

CHAPTER 1

CHAPTER 2

CHAPTER 3

CHAPTER 4

CHAPTER 5

CHAPTER 6

CHAPTER 7

CHAPTER 8

CHAPTER 9

CHAPTER 10

CHAPTER 11

CHAPTER 12

CHAPTER 13

CHAPTER 14

CHAPTER 15

List of Tables

Preface

One swallow does not make a summer, but one skein of geese,
cleaving the murk of a March thaw, is the spring.
(Leopold 1966)

Wild geese inspire us in many ways: they announce the seasons with their great flocks and migrations, they remain faithful to mates and family, and they amuse us with musical calls and fun antics. Aldo Leopold's (1966) *A Sand County Almanac* makes an important point: 'That's the question, I think, for our students, for ourselves, for all next generations. What if there be no goose music?' We have written this book about the inner workings of a Barnacle Goose society as a source of continued inspiration.

The lives of wild geese are extremely fascinating, not least because of the many problems and dilemmas that they encounter daily. Dilemmas are found in all areas of life, for example in choices regarding food, mates, migration, breeding, parental care, and neighbours in flocks. Individuals solve dilemmas in a variety of ways, and the collection of responses leads to the 'behaviour' of a population; accordingly a population may grow or decline in number and range. This exciting link between individual and population behaviour is an important theme in the book. Our model species is the Barnacle Goose, which we studied in two migratory flyways at locations in Britain, Norway (including Svalbard) and Sweden. The book was written for a broad readership, including those interested in bird biology in particular or wildlife ecology in general. Perhaps the 'unanswered questions' that we pose in some chapters will encourage others to choose geese and their habitats for study.

The production of an earlier version of this book (*Wild Goose Dilemmas*, 2007) was sponsored by the Directorate for Nature Management, Norway, the same agency that supported much of the field work in Helgeland through grants to the Wildfowl & Wetlands Trust. JMB acknowledges support from WWT Slimbridge and Caerlaverock; Scottish Natural Heritage; NINA, Department for Arctic Ecology, Tromsø; Large Animal Research Group, Department of Zoology, University of Cambridge; and Peter Scott Memorial PSTERIC Fund. JP was supported by the University of Groningen for many years. KL acknowledges support for the long-term studies in the Baltic from the Swedish Natural Science Research Council, the Swedish Environmental Protection Agency and Stiftelsen Olle Engkvist Byggmästare.

We thank the many ring-readers, goose-catchers, and all colleagues who had an important share in collecting data. We are grateful for the opportunity to collaborate with Des Callaghan, Sharmila Choudhury, Charlotte Deerenberg, Rudi Drent, Bart Ebbinge, Pär Forslund, Nils Gullestad, Lars Gustafsson, Martine Hausberger, Maarten Loonen, Annie Marshall, Carl Mitchell, Malcolm Ogilvie, Kees Oosterbeek, Myrfyn Owen, Paul Shimmings, Ingunn Tombre, Staffan Ulfstrand, Mennobart van Eerden, Henk van der Jeugd and Tom van Spanje. The facilities at WWT Caerlaverock to observe the geese are second to none and we thank the staff there for their continued investment. Studies on

Lånan, Norway, would have been impossible without the hospitality of the Johnson family. We gratefully acknowledge the long-standing, generous support of the Norwegian Polar Institute (Oslo, later Tromsø) and Governor of Svalbard (Sysselmannen).

For their contribution and encouragement toward the production of the book we are grateful to Gillian Black, Terje Bo, Kjell Einar Erikstad, Arild Espelien, Rob Fleischer, Larry Griffin, Mark Hulme, Derek Lee, Fridtjof Mehlum, Steve Percival, Eileen Rees, Paul Shimmings, Joost Tinbergen and Ingunn Tombre. Creating the book in its present form would have been impossible without the skills of Dick Visser. Earlier versions of chapters were read by a group of reviewers, and we benefited much from the expertise of Emily Bjerre, Mark Colwell, Götz Eichhorn, Luke George, Ken Griggs, Christiane Hübner, Ellen Kalmbach, Kees Koffijberg, Derek Lee, Anne Mini, Jeff Moore, Jan Allex de Roos, Henk van der Jeugd and Chris West.

We thank the publishers and societies for reproduction of previously published material, including Cooper Ornithological Society (*Condor*), Elsevier (*Animal Behaviour, Biological Conservation*), Oxford University Press (*Behavioral Ecology, Partnerships in Birds* (J. M. Black)), Norsk Polarinstitutt (*Skrifter*), Wiley/Blackwell/Academic Press (*Evolution, Functional Ecology, Ibis, Journal of Animal Ecology, Lifetime Reproduction* (I. Newton), *Molecular Ecology, Oikos, Ornis Scandinavica*). We also thank Mennobart van Eerden (for Figure 1.1), Joe Blossom and Myrfyn Owen (Figure 1.3), Pat Butler (Figure 2.7), and Addy de Jongh and John Videler (Figure 8.3).

CHAPTER 1
Introduction

WINTER IN SCOTLAND A thousand goose bodies flow steadily across green pastures. Suddenly, white cheeks flash as they stand erect when cows move towards the barn, causing a brief silence in the otherwise constant monotone of goose bills ripping leaves from the sward.

SPRING IN NORWAY At first light, small groups arrive on a northern sea-swept marsh. Travelling in pairs, females graze succulent new growth while males scan the horizon or chase adjacent flock members. The older females' bellies stretch toward the ground, packed with nutrient stores, ready to depart on the final stage of migration.

SUMMER ON SPITSBERGEN (SVALBARD) Goose nests fill barren islands, each pair defending a small patch of rocks. A group of females on break from incubation forage on tundra plants while mates guard the eggs.

AUTUMN ON BEAR ISLAND (SVALBARD) Goose pairs, some with goslings in tow, occupy the same patches as last year, aware of twists and turns in the landscape that hide either the best food or an approaching predator. The geese forage all day as darkness and cold will soon take over, forcing the southern retreat.

This description of life in a wild goose society sets the scene for our study about how individuals' decisions influence the growth and expansion of two Barnacle Goose *Branta leucopsis* populations. The Barnacle Goose in northern Europe resembles the smallest races of Canada Geese *Branta canadensis* in North America. It weighs about 2kg (*c.* 4.5lbs), lays up to seven eggs (typically four) and feeds on grasses and forbs. In this book we present results from long-term studies of a Norwegian high-Arctic population and of a population that began to nest on islands in the temperate Baltic Sea, 2,000km south-west of its original Arctic breeding grounds. Both populations migrate to spend winter months near rural communities that farm fertile soils reclaimed from adjacent coastal lowlands and estuaries.

We address two related questions in this book: what makes a successful goose and which individual characteristics drive population expansion? Our research agenda, therefore, is to describe the behavioural repertoire of wild Barnacle Geese, find out which individuals are most likely to survive and reproduce, and determine their effect on the demography and distribution of the population as a whole. The study is one of behavioural ecology and population ecology resulting from the processes of natural selection, including sexual selection and kin selection. Founded in our concern with the fate of this species, we comment on the bird's adaptable nature as it encounters rapid changes in its environment. The climate is changing particularly fast in northern areas. Increased temperatures affect the functioning of Arctic ecosystems (marine and terrestrial), ultimately determining what Barnacle Geese may achieve.

Individual variation and problem solving

Animals in the wild face one dilemma after another as they attempt to make ends meet. Throughout this book we use the term 'dilemma' to refer to a challenge or trade-off that must be resolved. Successful individuals must somehow assess the relative merits of behavioural choices as they strive to reproduce, thus passing on copies of genes in their offspring. As we began to identify the behavioural attributes of individuals in our study populations, we noticed great variability in performance and personalities (*sensu* Dingemanse & Réale 2005). For example, the population consists of a range of competitive abilities, from individuals that fight and win each encounter through to those that consistently submit and run away. Fighting can be expensive in terms of energy and risk of injury, but the pay-off of acquiring better food or mates may be worth the effort. On the other hand, some individuals acquire their share without fighting. The dilemma of whether to fight or not will therefore be decided by the relative costs and benefits involved and whether the outcome will better enable reproduction. Another way in which individuals vary has to do with timing, or organising their schedule of reproductive efforts within a season and over the course of a lifetime. We found that in a society of wild Barnacle Geese, as in other social vertebrates, an individual's choices are complex because the outcome often depends on what others do.

These and other problems geese face are in some ways similar to those of humans. Geese live in large groups, but they associate closely with just one partner in a pair bond that may last for life. Parents and offspring may coexist as a family until goslings are well beyond being able to fend for themselves. Geese can also be remarkable creatures of habit, travelling the same routes on journeys and visiting the same spots year after year to eat, sleep and nest. Yet when conditions change, some find a way to change and adapt to new opportunities. We consider how the extremes in long-lasting monogamy, prolonged biparental care and traditional use of sites might have evolved and describe the consequences of their maintenance in current-day goose societies.

Geese have a major eating problem. Their digestive system is rather inefficient at breaking down their food. To cope, these herbivores must pass large amounts of forage through their digestive tract to obtain the nutrients required for daily maintenance, let alone the extra nutrients that fill each egg in a clutch. Compared to other waterbirds, for example those that eat fish, geese must process much more food during a typical day. In the next section we discuss this physiological and morphological constraint.

The curse of vegetarian ancestors

Through the process of adaptive radiation, whereby different species evolve by making use of different resources, early waterfowl colonised a set of specialised niches along coasts and inland waterways; that is, wetlands (Owen & Black 1990). Figure 1.1 shows the position of Barnacle Geese among 15 other waterbirds that thrive in the coastal areas of Europe. Plants in wetlands are often relatively low in nutritive qualities or they contain secondary compounds that inhibit digestion (Buchsbaum *et al.* 1986, Gauthier & Bédard 1990). Ancestors of modern-day geese thrived by practicing unique styles of exploiting this type of food. To say that geese are vegetarians is an understatement. It is more appropriate to claim that geese are specialised grazing machines built with superior harvesting capabilities. The ancestors of each goose alive today survived to pass on their genetic material for harvest efficiency. Poor designs were weeded out, unable to obtain enough reserves to fuel costly flights and reproductive attempts. It was the successful ancestors that set the bounds for current goose behaviour.

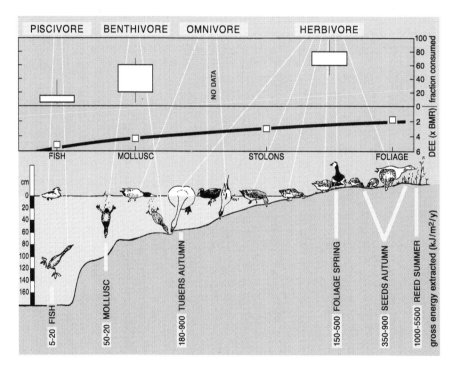

Figure 1.1. *Cross-section through a freshwater wetland showing food habits and habitat niches of a typical waterbird community in Europe. The herbivorous Barnacle Goose, fifth bird from the right, is a key herbivore in this system, exploiting a zone of coastal vegetation. Other species specialise in a variety of other locations and food types. The figure shows that the herbivores (i.e. geese) have a low-cost lifestyle and yet harvest the largest quantity of food, whereas fish eaters, having a high-cost feeding mode, have the lowest levels of food exploitation in the wild. The approximated cost of foraging (middle panel, Daily Existence Energy DEE expressed as multiples of Basal Metabolic Rate BMR) and the gross energetic gain extracted from the food taken (lower panel; kJ per m² per year) are indicated together with harvest levels (upper panel, fraction of available food consumed). From van Eerden (1997).*

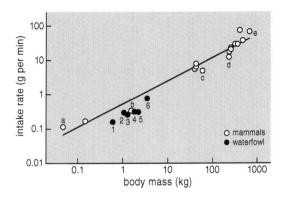

Figure 1.2. *Comparison of food intake rates among herbivores when grazing on grasses. The food intake rate in herbivorous waterfowl is related to body mass though they ingest less food per time feeding than mammalian herbivores of the same size.[1] Selection for high-quality food is one way to compensate for a lower food intake. Note the log-scale of the axes; the exponent relating the intake rate to body mass in mammals and waterfowl is 0.69 (SE 0.034), corresponding to the allometry of energy requirements. The waterfowl species are (1) Wigeon Anas penelope, (2) Red-breasted Goose Branta ruficollis, (3) Dark-bellied Brent Goose Branta bernicla bernicla, (4) Barnacle Goose, (5) Pink-footed Goose Anser brachyrhynchus, (6) Greylag Goose Anser anser; from Prop & Deerenberg (1991), Therkildsen & Madsen (2000), Durant et al. (2003), Prop & Quinn (2004). The mammals presented cover the range from (a) lemmings, (b) rabbit, (c) goat, (d) elk, through (e) bison; data from Shipley et al. (1999) and Iason et al. (2002).*

Today's 15 species of geese exploit a broad range of plant communities. By plotting the speed with which each species can harvest food against bill size, Owen (1980a) showed how the smaller species, with stout nimble bills, are able to peck at extremely high rates when harvesting short saltmarsh plants. The larger geese, with long bills that have serrated lamellae, forage at a slower rate, and are better at digging for tubers underground and stripping seeds from tall stocks above the ground. Geese do not achieve the high food intake rates of mammalian herbivores, which are able to take much larger bites (Figure 1.2), but compensate by increasing pecking speed and focusing foraging effort on high-quality foods.

While geese may be effective harvesters, they are relatively ineffective at processing food once it is obtained. Instead of having complex digestive systems, as in mammalian herbivores, geese are equipped with an expandable oesophagus, a gizzard, a slender intestine, and two caeca (Figure 1.3). As a consequence, they largely depend on the juicy parts of plant cells as a source of energy and nutrients, whereas the tougher cell walls, which usually make up more than 60–75% of the biomass, are only partially digested. Usually much of the food that is eaten passes through the gut in an undigested form and about 70% comes out the other end. Geese cope with a poor digestive capability in two important ways. First, they are highly selective, taking the most nutritious food items, and second, they are extremely dedicated foragers. For example, when feeding during the daylight period only, Barnacle Geese forage for seven hours of an eight-hour winter day. In spring, when daylight allows they may forage for 17 hours or more, ingesting an amount of food that is close to their body mass. Thus, they process large amounts of food that are harvested at incredibly fast rates.

Geese possess one characteristic that is perhaps the key to their evolutionary success – their ability to select the food they need. When harvesting the sward with an impressive rate of up to 300 bites per minute they are able to take the most nutritious parts of the grasses and avoid

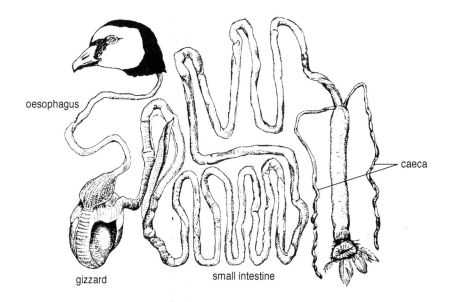

oesophagus

caeca

gizzard small intestine

Figure 1.3. *Outline of the alimentary tract of a Barnacle Goose. The oesophagus is a functional 'crop', which is most effectively used in summer. The geese fill their oesophagus during brief incubation breaks and return to the nest to digest the food while incubating the eggs. Grit particles are collected in the gizzard where cell walls of food plants are broken into small fragments, thus facilitating absorption by the gut. The caeca are sites of intensive microbial activity. There are cyclical changes in gut size through the annual cycle depending on the birds' diet and needs for drawing nutrients from the gut. After Owen (1980a).*

dead and older leaves (Owen 1978, 1981a, Fox 1993, Kristiansen *et al.* 2000). Apart from this micro-selection, geese must cope with the non-uniform distribution of their preferred food. Habitats on which Barnacle Geese depend for most of the year comprise a vegetative mosaic that differs in availability of nutritious plants. Because of the constraints imposed by their poor digestive system it is crucial for the geese to find those patches and zones that provide the most nutritious food items. Seemingly simple tasks of deciding which food to choose, how fast to eat it and when to move on are essential elements of a successful goose.

Of course, much like any other habitat, wetlands are also home to secondary consumers, or predators. Therefore, the ancestral goose also had to be adept at avoiding predation. A common theme throughout our study, therefore, is the description of strategies Barnacle Geese use to optimise foraging performance while avoiding predators.

Behavioural ecology and conservation management

Behavioural ecologists suggest animals are involved in 'decision making' or 'problem solving' and attempt to identify potential 'rules' and 'cues' that animals use to process information. Such terminology does not mean to imply a cognitive process of solving complicated calculations. Rather the approach is a way to understand which factors have importance in shaping the animals' performance during the process of evolution (Davies *et al.* 2012). Understanding why only certain individuals survive and reproduce is central to describing natural selection in action, as well as providing insight into the 'behaviour of populations'. We believe that caretakers of wildlife populations armed with this individual animal approach to understanding populations are better equipped for directing conservation and management initiatives (see Clemmons & Buchholz 1997, Caro 1999, Gosling & Sutherland 2000, Festa-Bianchet & Apollonio 2003).

It is a challenge to describe how the behaviours of individuals influence populations (Goss-Custard 1996, Sutherland 1996). Most wildlife biologists are faced with the difficulty of determining where their animals go once they leave the study area. Even when working on closed populations that are confined to life on an island, a lot of hard work is required to catch, mark and follow individuals with enough stealth and effort to determine where they go, who they meet, what they eat and how they fare. To conduct such studies in long-lived, migratory geese is doubly challenging.

We were able to study the population ecology of Barnacle Geese because of their remarkably consistent use of the landscape. Like clockwork, our Arctic study population arrived at the same stretches of coastline in Scotland and England (in autumn), mid-Norway (in spring) and Svalbard (in summer). In the early days of our study, the population was small enough that nearly all individuals in our marked sample were resighted several times per year. It was relatively easy to keep 15–25% of the population marked with individual leg-rings (bands) by catching the geese when they replaced their flight feathers during the moult in summer. The biggest problem was getting up to the Arctic with all the gear. All credit goes to our mentors, Myrfyn Owen and Rudi Drent, for having the foresight to launch this long-term study. Their visionary planning and hard work provided the foundation for the pursuit of describing the population consequences of individuals' behavioural decisions.

Parallel to the studies of the Arctic population, we also performed long-term studies of the temperate-breeding population in the Baltic Sea region. Several aspects of behaviour, reproductive success and population expansion were therefore analysed in detail in both Barnacle Goose populations. Although the focus of our research was slightly different in the two populations, the basic research methodology was similar.

Goose research

When we began, goose biology was a growing specialty field, rich with initial information and many unanswered questions. The foundation for our work was laid by the efforts of no fewer than 50 key workers that studied various goose populations during the 30 years prior to our study. In particular, the detailed observations of the social organisation of White-fronted Geese *Anser albifrons* (Boyd 1953) and Canada Geese (Collias & Jahn 1959) elevated the goose as a worthy model system. With the publication of his book, *Wild Geese of the World*, Owen (1980a) influenced a new generation of goose biologists. The book reviewed the current state of knowledge and proposed fertile avenues for future research in ecology, evolution and management of goose populations. While initial studies may have been driven by questions involved in 'managing the resource', it soon became evident that geese were ideal subjects to pursue current theoretic issues in ecology and evolution. Our goal was to marry these approaches. The ability to rigorously manage populations, either for the enjoyment of human users or for the animals in their own right, depends on our understanding of individuals' behavioural adaptations (see Owen & Black 1990).

Dennis Raveling's ideas about the causes and consequences of individual goose types (or personalities as we might call them now) that make up populations were particularly insightful. His paper, published in *Journal of Wildlife Management,* describes how only a small proportion of the Canada Goose population under study succeeded in reproduction and that individuals were consistently successful or unsuccessful over time (Raveling 1981). He argued that variation in reproductive success might be determined by an individual's ability to fight and acquire resources (food, mates, territories), where only dominant birds were successful breeders, creating a highly skewed distribution of success in the population. He hypothesised that less able fighters were only successful when environmental conditions changed in their favour and that such changes, while currently rare, may have been the norm in the past. He was describing a theory referred to as an 'evolutionary stable strategy' where, for example, over time both forms of the fight or no-fight strategy may enjoy equal success (Maynard Smith 1982). Another set of Raveling's papers described how some individual Canada Geese tended to return to the same locations across seasons and years (Raveling 1970, 1978, 1979). He suggested that fidelity to particular sites might enable family members to reunite and that maintaining social units may result in clusters of individuals or subgroups in the population. Raveling's ideas stimulated and directed a major line of inquiry in this book: starting with the development of individual aggressiveness in goslings, to the establishment of pair bonds, finding food, acquiring nutrient reserves, establishing territories and maintaining a traditional use of the landscape, and identifying the survival and reproductive success of various goose types (dominant through subordinate, site faithful through pioneer, etc.) and their contribution to future generations.

Individual differences (plumage and vocalisations)

One of the most prominent plumage features of a Barnacle Goose is its white cheek-patch, which extends across the forehead and under the lower mandible and contrasts with the dark black neck, black eye-patches and black bill. As is the case with Canada Geese, the white cheek-patch is used for signalling information to mates, kin and flock members. By waving the head in the air or stretching it forward or to the left or right a goose apparently indicates its intentions in aggressive encounters and other social displays (Radesäter 1974, Black & Owen 1988). Pair and family members coordinate movement through the flock or departure from the flock with preflight head-tossing movements (Black & Barrow 1985). These preflight signals, which make use of the striking white cheek-patch, are performed to facilitate synchronised take-off, thus maintaining the integrity of family units (Raveling 1969a, Black 1988).

Individuals vary to some extent in plumage colour on their wings and bodies, but we suspect the plumage pattern on the face is one of the features by which individuals recognise each other, in addition to vocalisations. Variation in the shape and brilliance of the white cheek-patch, together with the different sized eye-patches, makes each individual distinct (Figure 1.4). It is possible even for us to recognise differences among some individuals, though we are not nearly as adept as some swan biologists who readily identify hundreds of individuals with distinctive face patterns (Rees 2006). Avid birders that scan the black, white

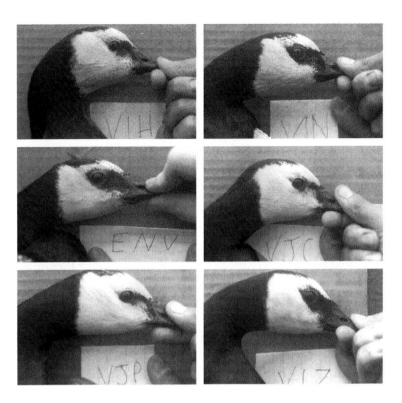

Figure 1.4. *Face patterns of six wild-caught Barnacle Geese (females left, males right). Nearly all individuals have distinctive face patterns, varying in the amount of white above the eye and black above the bill. In Chapter 4 we show that females with darker faces sample more potential mates before settling on a long-term partner, whereas lighter-faced females partnered up without sampling as many alternatives (Choudhury & Black 1993). Determining the role of face patterns and other plumage ornaments in waterfowl, whether for individual recognition or status/quality signals, is fertile ground for students interested in sexual selection. Photo by Sharmila Choudhury.*

and grey bodies in dense Barnacle Goose flocks may also detect subtle variations. In the Svalbard population a few 'white' Barnacle Geese stand out from the rest. These individuals have a recessive gene expressing 'leucistic' plumage without melanin pigment, but still with normal eye colour, black legs and bills (Owen & Shimmings 1992). These birds behave and pair normally but if the mate also carries the gene it is expressed in some or all of the goslings in their brood. The first white bird was recorded in the Svalbard population in the 1920s and one or more have been seen in most years since then; four 'white' birds were present in recent winters, including a family of three (Griffin 2013).

Barnacle Goose loud calls, which are higher in pitch than the honks of the ubiquitous 'honker' (large Canada Geese), are also individually distinct (Figure 1.5). Calls appear to be used to monitor the whereabouts of partners and family members; when birds become separated from family members within large foraging flocks they stand erect and march through the flock, calling loudly. Reunion is swift when the mate answers with their call. In flight some pair members and offspring call continually, presumably to ensure that they do not become separated among the masses of others.

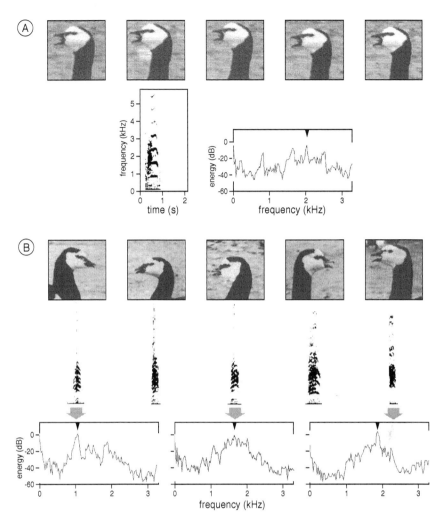

Figure 1.5. *Variation in individual loud calls in Barnacle Geese: (A) Five calls from one individual, (B) calls from five different individuals. The amplitudes (pitch/tone) of the calls were readily quantified using sonographic software (from Hausberger et al. 1991). Calls were consistent within individuals but quite distinct among individuals (see also Hausberger et al. 1994). Discovering the role vocalisations play in the social lives of waterfowl will be an exciting topic for future students. Photos by Martine Hausberger.*

Svalbard Barnacle Goose population

Our primary study population was surveyed each winter on the Solway Firth, a large estuary where several rivers meet the Atlantic at the westernmost border of northern England and southern Scotland. Concern for the population began back in the winter of 1948, when only 300 individuals were found (Owen & Norderhaug 1977). The Barnacle Goose was legitimate quarry throughout its range in the United Kingdom and Norway and was heavily shot in the late 19th and early 20th centuries (Harrison 1974). Shooting during the Second World War was particularly severe on the Solway Firth; goose roosts on the saltmarsh and mudflats were used as artillery and naval firing ranges. Exploitation in Svalbard during the breeding season was probably intense prior to protection in 1955. Fishermen, whalers, trappers, coal miners and egg collectors were all reported to have taken eggs, down and birds. One British 'nesting expedition' to Svalbard in 1931 (15 June – 5 July) searched the west coast of Spitsbergen for eggs, and collected clutches from the nests of 23 Light-bellied Brent Geese *Branta bernicla hrota*, 10 Pink-footed Geese *Anser brachyrhynchus* and one Barnacle Goose (Lings 1935). The Barnacle Goose was apparently the rarest and their eggs were most sought after. The collectors saw just 12 Barnacle Geese feeding at North Fjord, one of the branches of the large Ice Fjord. Similar egg collecting expeditions that focused on the prized Barnacle Goose were reported in the American Ornithologists' Union journal, *The Auk* (Jourdain 1922).

After egg collecting and hunting were banned, and breeding sanctuaries and winter refuges were established, the Svalbard Barnacle Goose population increased to unprecedented levels (Figure 1.6), making this one of the most noteworthy wildlife management success stories in Europe (Norderhaug 1984). The population size increased to 5,100 individuals in the winter of 1973, 10,500 individuals in 1984, 26,100 individuals in 2002 and 31,000 in 2012. As the number of geese increased, so did the number and size of breeding colonies, causing an increase in competition for food and nest locations. The number of colonies increased from just five prior to 1960, to 11 in the 1960s, 37 in the 1980s and about 50 in the 1990s (Mehlum 1998). Today it is hard to keep track of the numerous colonies, which occur at suitable locations throughout the archipelago (Figure 1.7). Most of the colonies are found along the west coast and in the valleys of the large fjords, but many islands in the east are occupied by breeding Barnacle Geese as well (Goosemap 2013).

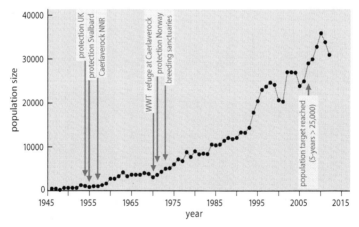

Figure 1.6. *Svalbard Barnacle Goose population size and management actions. Population management began with protection from hunting in the mid-1950s. Then in 1957 a large portion of the winter saltmarsh habitat was turned into a National Nature Reserve (at Caerlaverock) and this included some zones that prohibited the take of any waterfowl so the geese could feed undisturbed. Numbers grew to 3,700 in 1965. After The Wildfowl & Wetlands Trust (WWT) purchased and began managing 600ha of pastures and saltmarsh for the geese (in 1970) and 'no-go zone' Bird Sanctuaries were established in Svalbard (1973), the population rose to an unprecedented total of 8,800 in 1978. During our study, the population was fully protected and it continued to grow. In 2007, numbers reached the conservation target of 25,000 for a five-year period. Values updated from Owen & Norderhaug (1977) and Owen & Black (1999) with counts by Paul Shimmings and Larry Griffin (WWT).*

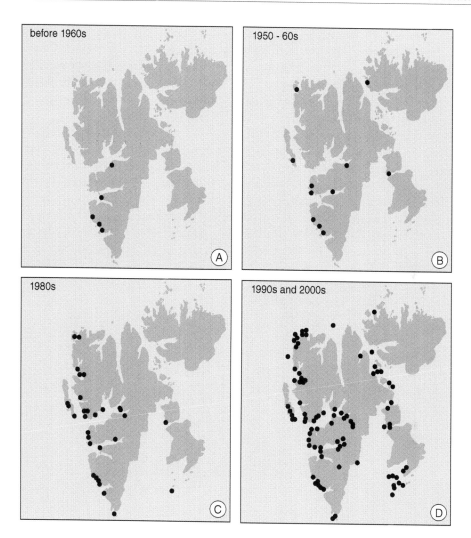

Figure 1.7. *Distribution of Barnacle Goose colonies in Svalbard (A) before 1960, (B) during the 1950s–1960s, (C) 1980s and (D) 1990s and 2000s. During colony surveys coastal sites were approached by boat when weather and pack ice conditions would allow. Nests were found on flat offshore islands, steep cliffs and in canyons. The number and density of nests also increased at each of these colonies during the study. Panel A–C from Mehlum (1998), panel D after Goosemap (2013).*

The recovery of this population is comparable to other successful initiatives to save dangerously small populations like, for example, those of Peregrine Falcons *Falco peregrinus* and Aleutian Cackling Geese *Branta hutchinsii leucopareia* in North America (White *et al.* 2002, Mini *et al.* 2011). Coincident with the increasing size of the population, flocks began to overflow traditional boundaries and colonise new agricultural areas where the geese are often not welcome. Now wildlife managers are blessed with the dilemma of designing policies that satisfy those who view the once endangered population as a pest and those that celebrate its return. Many goose populations are actively harvested by hunters and gatherers that thrive on the sport and savour the taste. In fact, much of wild goose research has been accomplished with the aim of enabling safe hunting quotas while ensuring healthy-sized populations that can withstand future harvesting (Trost *et al.* 1993). The Barnacle Goose was fully protected during most of our study, giving us an extraordinary opportunity to observe and describe a population that is close to being 'natural'. Furthermore, the population was initially small enough that we were able to reliably estimate numbers on an annual basis and to keep track of birds that strayed to other populations.

Figure 1.8A shows the composition of the steadily increasing Svalbard Barnacle Goose population between 1959 and 2012, including the number of juveniles and yearlings in the wintering population and the number of successful and unsuccessful (i.e. non- or failed) breeders. Both successful and unsuccessful breeders increased in numbers, but, somewhat surprisingly, the unsuccessful breeders were quickly outpacing the successful breeders. This means that it became harder and harder for the average goose to actually succeed in reproduction, as also reflected by a decreasing probability of breeding successfully during the study (Figure 1.8B). It is remarkable that when Myrfyn Owen initiated the Svalbard Barnacle Goose study in 1973, the 1,500 breeding-age pairs returned to the wintering grounds

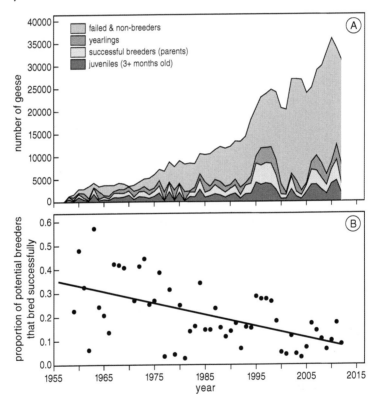

Figure 1.8. *(A) Composition of the Svalbard Barnacle Goose population, including number of juveniles in the autumn flocks, number of successful breeders and non-breeders (including failed breeders). Most notable is the huge increase in failed and non-breeders. (B) Proportion of potential breeders that returned to the wintering grounds with at least one surviving gosling.[2] Barnacle Geese become potential breeders in their second year of life. Apparently it was more challenging for individuals to hatch and rear goslings as competition for food and space intensified when the population increased in size. Formulas for calculating these lifetable parameters are described in Chapter 14. Updated from Owen (1984), Pettifor et al. (1998).*

with 1,070 juveniles, whereas in 2004, when the population was much larger, 12,670 pairs produced only 590 young. By 1986 we calculated that out of every 100 potential eggs (25 mature females × clutch size of four), only eight eggs resulted in fledged young reaching the wintering grounds (Owen & Black 1989a). Relating details of this decline in reproductive success, which is referred to as a 'density-dependent effect,' and describing how the population still managed to increase, are primary aims in this book.

Baltic Barnacle Goose population

Our second Barnacle Goose population was studied in varying detail since its natural establishment in 1971 in the Baltic Sea region. The founders of the Baltic population most likely originated from the Arctic Russian population, which made use of Baltic habitats as a spring staging area on their way to their Arctic breeding ground (Larsson *et al.* 1988, Larsson & van der Jeugd 1998, Ganter *et al.* 1999). The Baltic population grew rapidly from only one breeding pair in the summer of 1971 to 17,000 individuals in 1997 and to more than 40,000 individuals in 2012. The first breeding colony was established on three small, closely situated islands, named Laus holmar, off the east coast of the larger island of Gotland, Sweden, in the central Baltic Sea. In the beginning of the 1980s Barnacle Geese also started to breed on other islands off the coast of Gotland, the west coast of Estonia and the east coast of Öland, Sweden. The range expansion within the Baltic Sea region continued in the 1990s and 2000s, and in 2012 colonies of varying sizes were found along the coasts of mainland Sweden, Finland, Estonia and Denmark. Throughout this book the 'Baltic population' refers to Barnacle Geese breeding in the total Baltic Sea region and the 'Gotland colonies' refers to geese breeding on small islands off the coast of Gotland. In 1997 about 75% of the breeding pairs in the Baltic population bred on Gotland, whereas in 2012 – that is, after the range expansion within the Baltic region – this proportion had declined to about 20%.

During winter, the Baltic population mixes with the much larger Arctic Russian and North Sea Barnacle Goose populations on the wintering grounds in the Netherlands and Germany. From resightings in northern Russia of marked Barnacle Geese from the Gotland colonies we know that some exchange of individuals between the Russian and Baltic populations is occurring. However, the observed rapid growth and expansion of the Baltic population can be largely explained by high reproductive output and low adult mortality in the Baltic (see Chapter 14). As the number of breeding pairs increased in the larger Gotland colonies we detected strong density-dependent effects on the survival of newly hatched chicks, leading to a dramatic decrease of the production of fledged young per breeding pair.

Like the Svalbard population, the Baltic population feeds on agricultural fields in increasing amounts. Some hunting has therefore been allowed in Sweden in spring and autumn, and in the Netherlands and Germany in winter, under a special licence to protect crops from damage. This limited hunting is not likely to have an impact on the Baltic population size. However, Holm & Madsen (2013) found that 13% of the Baltic/Russian Barnacle Geese carried shotgun pellets, suggesting a considerable mortality rate due to hunting. Much of the shooting might occur in Russia, where Barnacle Geese are allowed to be hunted during spring and autumn migration, thus probably not affecting the Baltic population. The main study colonies on Gotland are protected during the summer.

Layout of the book

After the first three introductory chapters the remaining chapters can be described under several broad themes (Figure 1.9), including chapters on life history decisions (Chapters 4–7), daily decisions (Chapters 8, 11) and annual decisions (Chapters 12, 13). These decisions result in two levels of consequences, including those for the individual (Chapters 9, 10) and, finally, those for the population (Chapters 14, 15). In this section we briefly outline each of the chapters and provide examples of the types of dilemmas involved.

In Chapter 2 we present additional background information about our two study populations and study areas. We also develop ideas about the historical distribution of the species during the last ice age. Chapter 3 provides details about the field techniques we used to study the birds and their food plants. Chapter 4 is about the geese sampling potential mates in a series of 'trial liaisons', which ultimately results in a long-term, monogamous

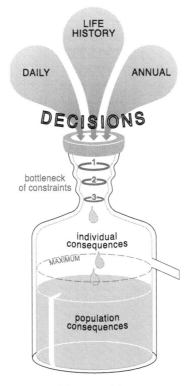

Figure 1.9. *Diagrammatic representation of the types of decisions geese make and the consequences of those decisions with regard to the individual bird and the population as a whole. Daily decisions include those made on an hourly basis. Seasonal decisions are those made during the annual cycle. Life history decisions include events like parent–offspring relationships, choice of mate, when to initiate a breeding career, production of different clutch sizes, etc. Note the decisions are forced through the 'bottleneck' of constraints that are imposed by the birds' (i) morphology, (ii) physiology and (iii) their environment (including, for example, interaction with flock and colony members). Consequences for the individual will be determined by resources a goose can acquire, which will ultimately determine its body size and fitness (survival and reproduction). Consequences for the population will be determined by the individuals that manage to survive and reproduce.*

partnership. We quantify reproductive success of different types of partnerships in terms of the partners' ages, experience and body sizes. In Chapter 5 we show how persistence in pair bonds, year after year, results in the production of more offspring and identify some of the unusual conditions when divorce occurs. In Chapter 6 we assess the relationship between goslings' early life experiences and their future survival and breeding prospects. We consider two dilemmas, one between parents and offspring and another among siblings within families, and quantify the pay-off for goslings that seek a prolonged association with parents. Chapter 7 deals with some of the exceptions to the usual goose lifestyle beyond the social pair bond, including dumping eggs in another's nest, adoption of goslings, and formation of groups of pairs that form clusters in space and time. In Chapter 8 we describe behavioural tactics that geese employ while foraging. We investigate how geese solve the problem of finding food in a landscape influenced by topography, climate and competition with other geese. We ask whether foraging performance is related to dominance status and knowledge of feeding areas. In Chapter 9 we describe the probability of survival over the birds' lifespan and the pattern of reproductive success that can be achieved. We explore the relationship between breeding and subsequent survival with the view that breeding can be dangerous and energetically expensive. We explain the consequences of initiating breeding early and late in life. In Chapter 10 we examine factors that influence growth and ultimate body size. We describe how body sizes are shaped in the first weeks of life by prevailing environmental conditions, including food quality and population density. Chapter 11 is about the timing of successive events in the annual cycle. Geese must solve the dilemma of foraging enough before beginning the long-distance migration and arriving early enough to hatch and rear goslings in the narrow window of time that is available in short Arctic summers. In Chapter 12 we consider the goose dilemma of site fidelity or dispersal and how this choice is influenced by an individual's characteristics as well as by the choices of others. Limitation of food resources results in intense competition among flock and colony members, which is thought to stimulate processes like colonising new breeding areas or expansion of the winter and spring range. Chapter 13 investigates the occurrence of immigration and emigration among Barnacle Goose flyways. Chapter 14 is about population dynamics, including information on births and deaths. We describe the likely mechanisms behind the populations' continued growth and periods of stability from the perspective of density-dependent effects and other factors. We discuss implications for colonial nesting geese that must cope with increasing predation from Polar Bears *Ursus maritimus* and White-tailed Eagles *Haliaeetus albicilla*. Chapter 15 is about the flexible and resourceful nature of wild geese and how they have come to rely on food plants found in lush agricultural pastures. We discuss some of the issues involved in balancing the needs of rural farming communities and wild goose populations.

Summary

This book focuses on a population of Barnacle Geese that breeds in the high Arctic in Svalbard, Norway, and another that colonised and now breeds at its spring migratory stopover location in the Baltic Sea region. We ask two related questions in this book: what makes a successful goose and which individual characteristics drive population expansion?

The study is one of behavioural ecology and population ecology resulting from the processes of natural selection (including sexual selection and kin selection). We believe that wildlife managers equipped with an evolutionary perspective that explains why individuals behave in a certain way are better able to design effective management and conservation actions.

Each of these grey, black and white geese are individually different due to unique plumage patterns on their faces and distinctive vocalisation characteristics. The recovery of the Svalbard Barnacle Goose population is one of the most noteworthy wildlife management success stories in Europe. Numbers rose from a low of 300 birds in 1948 to a level of 31,000 in 2012. The population responded to protective measures and habitat management initiatives in a stepwise fashion with brief periods of stability and rapid growth. As the population grew, a decreasing proportion of the geese managed to produce offspring. In the early years as much as 60% of the breeding population succeeded in bringing offspring back to the wintering grounds, whereas in more recent years only 20–30% has been able to do so. The development of our second study population, the naturally established Baltic population, was even more striking than the Svalbard population. These colonies, which most probably originated from the Arctic Russian population, increased from one breeding pair in 1971 to more than 40,000 individuals in 2012. As in the Svalbard population, strong density-dependent effects on reproduction were found. The aim of the book is to describe the ways in which geese are able to survive and breed successfully, thus fuelling the continued growth and expansion of these populations. Our goal is to make a contribution toward describing population consequences of individuals' behavioural decisions.

Statistical analyses

[1] Food intake rate in herbivorous waterfowl was related to body mass ($F_{1,4} = 23.9$, $P < 0.01$), though they ingest less food per time feeding than mammalian herbivores of the same size ($F_{1,19} = 14.1$, $P = 0.001$).

[2] The proportion of potential breeders that succeeded in producing surviving goslings declined over the course of the study (Pearson correlation coefficient $r = -0.569$, $n = 53$, $P < 0.0001$). Updated from Owen (1984), Pettifor *et al.* (1998).

CHAPTER 2

Study populations and study sites

In this chapter we describe our two study populations, the Svalbard and the Baltic populations, in relation to the global Barnacle Goose population. After providing information about the history of the populations, the chapter is organised according to the annual cycle for the Svalbard birds that travel to the high Arctic. We studied the Svalbard population at all parts of their migratory range, including breeding areas (Svalbard), migratory staging areas (coastal Norway), and wintering areas (Scotland/England). We focused on the Baltic population at the newly established breeding colonies on the island of Gotland, Sweden, in the central Baltic Sea. We provide a description of the birds' habitats and main predators in these areas. The chapter ends with a discussion about the ancestral origin of Barnacle Geese in relation to periods of global glaciation, which must have given rise to the birds' notable ability to track the occurrence of food plants.

Barnacle Geese

The species is known by a variety of names in Europe (Table 2.1), including the 'white-cheeked goose' in Germany, Norway, Sweden, Finland and Estonia, and the 'nun goose' in France and Germany, apparently because of the similarity with a black-and-white headdress. The English name for 'Barnacle Goose' came from early speculation about the origins of sea-going fowl that was linked to an arthropod/crustacean, the barnacle. Loonen (2005) describes the account from the 1600s: 'It was said that some trees in Scotland grew curious fruits. When these fruits were ripe, they fell into the water and passed through a barnacle-like phase before returning in autumn as fully developed geese.' Owen (1990a) suggested:

Language	Common name	Explanation
English	barnacle goose	named after the arthropod
Danish	bramgås	black goose
Dutch	brandgans	ambiguous, uncertain
Estonian	valgepõsk-lagle	white-cheeked goose
Finnish	valkoposkihanhi	white-cheeked goose
French	bernache nonnette	nun goose
German	weißwangengans	white-cheeked goose
German	nonnengans	nun goose
Icelandic	helsingi	goose with neck band
Irish	gé ghiúrainn	barnacle goose
Norwegian	hvitkinngås	white-cheeked goose
Portuguese	ganso-de-faces-brancas	white-faced goose
Russian	beloschokaya kazarka	black goose white cheeks
Scientific	*Branta leucopsis*	white-faced dark goose
Spanish	barnacla cariblanca	white-faced dark goose
Swedish	vitkindad gås	white-cheeked goose

Table 2.1. *Common names for the Barnacle Goose explained.*

'The legend did, no doubt, persist longer than it might otherwise have done because the geese were considered to be fish and could therefore be eaten by Catholics on Fridays.'

The Western Palaearctic supports five Barnacle Goose populations – three others in addition to our two study populations. Their names are based on the location of the breeding areas. There are three Arctic and two temperate breeding populations (Figure 2.1): these include the Greenland population which winters in Ireland and western Scotland; the Svalbard population which winters in northern Britain; and the Russian, Baltic and North Sea populations which winter mainly on the North Sea coast in Germany and the Netherlands (Madsen *et al.* 1999, van Roomen *et al.* 2003). Recent winter estimates for these populations are: Svalbard 31,000 in 2012 (WWT 2013), Greenland 80,670 in 2012 (WWT 2013), and Russian/Baltic/North Sea 770,000 in 2008 (Fox *et al.* 2010), with about 100,000 of these attributed to the Baltic and North Sea. This amounts to a number approaching a million individuals for the species. Resightings of engraved leg-rings indicate that some birds move among the populations, but in many cases wayward birds return to 'home' populations (Chapter 13). Although there is no recognised morphological variation among the Barnacle Goose populations, analyses using molecular techniques have revealed measureable genetic differentiation. Least genetic differentiation was found between the Russian and Baltic populations (Jonker *et al.* 2013).

Our two study populations exhibit two contrasting modes of living. The Svalbard population faithfully travels between the Arctic breeding and temperate wintering areas, covering a distance of 3,100km for each leg of the journey. The migration distance for birds breeding on Gotland in the central Baltic Sea, on the other hand, is only 750km from the

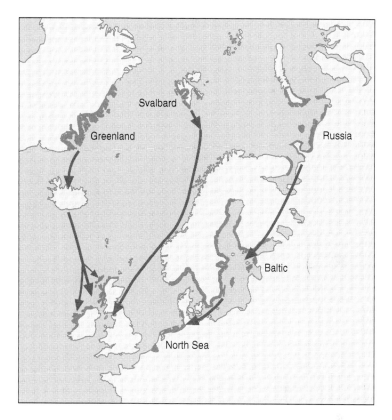

Figure 2.1. *Range of the five Barnacle Goose populations in Greenland, Svalbard, Russia, the Baltic and along the North Sea coast. The Russian, Baltic and North Sea populations use the same wintering areas in Germany and the Netherlands.*

wintering area. The fact that the Gotland Barnacle Geese now breed in habitats that are used during spring migration by Arctic Russian birds attests to the flexible nature of the species. This adaptability of lifestyles is also apparent for the newer North Sea Barnacle Goose population, which in 1982 began breeding even further south on coastal sites around the sea, lakes, rivers or embankments within the wintering range (Feige *et al.* 2008). For these North Sea birds the traditional long-distance migration between separate breeding and wintering sites was abandoned altogether. This behavioural change led to adjustments of several additional life history traits (Feige *et al.* 2008, van der Jeugd *et al.* 2009, Jonker *et al.* 2011, 2013, van der Jeugd 2013).

History of the Svalbard population

The discovery of the Svalbard archipelago marks the beginning of our knowledge of Arctic goose migration in general. In 1596, the authorities of Amsterdam sent two ships to find a new route to China and the Indies to enhance the economy. The expedition was dramatic, with several of the crew dying, including Commander Willem Barentsz. In the eyes of the

Amsterdam traders the expedition was probably a failure, but scientifically the enterprise must be regarded differently. On their way north they discovered Bear Island, which was given this name after a frightening encounter with a Polar Bear (de Veer 1598). Ten days later they were the first men in historical times to see the 'land with peaked mountains', or Spitsbergen. On landing, Barentsz and his men were surprised to find a colony of Brent Geese, as they were familiar with the species but did not know where these geese bred. Rather than the scientific satisfaction of this discovery the crew probably enjoyed more the taste of the eggs they brought back to the ship. Today the name Spitsbergen is reserved for the western islands, whereas the name Svalbard indicates the whole archipelago including the islands in the east, like Edgeøya and Barentsøya, and Bjørnøya (Bear Island) in the south.

Barnacle Geese were discovered in Svalbard much later. They were first noted in the scientific literature in the 19th century (Løvenskiold 1964, Norderhaug 1984). That Barnacle Geese were not seen in Svalbard is remarkable because 8,000–10,000 apparently wintered on Solway Firth estuary (our current study area) in the 1880–90s (H. Boyd, pers. comm.). Bio-archaeological studies of 17th and 18th century settlements in Svalbard have never provided evidence that Barnacle Geese, or their eggs, were eaten in former days (L. Hacquebord, pers. comm.). This is different to findings in Greenland, where Barnacle Goose remains from former times are common (Bennike *et al.* 1999). Owen & Shimmings (1992) suggested that the Svalbard population was established only a few centuries ago, speculating that the individuals founding the new population might have originated from the Russian population. Although we cannot exclude the possibility that Barnacle Geese colonised Svalbard after the 19th century, it seems equally plausible that the breeding population was too small to discover.

There is some speculation about the origin of Barnacle Geese on the Solway Firth in the early 1900s. It is suspected that this wintering area was shared by geese coming from Greenland and Svalbard (Hugh Boyd, pers. comm.), but the relative composition of flocks was unknown. After the Second World War when Barnacle Goose numbers on the Solway Firth declined to just a few hundred birds, there was great interest in finding out the origins of the remaining birds. Encouraged by the success of a group of students from Oslo University, who captured 685 moulting Barnacle Geese in Svalbard in the summer of 1962 (Bang *et al.* 1963), The Wildfowl Trust used rocket-propelled nets to capture 316 geese in Scotland in February 1963. This second catch included 94 of the birds ringed in Svalbard. This was the first solid evidence of the Norwegian–Scottish connection after the less substantial ring recaptures in the 1950s (Løvenskiold 1964). That none of the more than 1,000 metal rings previously fitted on Greenland and Russian Barnacle Geese were recaptured on the Solway Firth also supported the notion that the small wintering population came from Svalbard (Boyd 1964).

History of the Baltic population

In contrast to the Svalbard population, the first breeding attempts on Gotland and the further rapid increase and range expansion within the Baltic region are well documented (Larsson *et al.* 1988, Leito 1996, 2011, Larsson & van der Jeugd 1998, Feige *et al.* 2008, Mortensen 2011, Väänänen *et al.* 2010, 2011). All populations of Barnacle Geese used to be known as typical Arctic breeders; that is, until breeding pairs appeared in 1971 in the Baltic Sea some 2,000km from the nearest Arctic breeding area known at that time. The open coastal habitats around the large Baltic islands of Gotland and Öland, Sweden, and of Saaremaa and Hiiumaa, Estonia, have been used by migrating Russian Barnacle Geese as a spring staging area during at least the past century, and probably much earlier (Wibeck 1946, Kumari 1971). By 'short-stopping' to breed on Gotland and on other islands in the 'spring staging area' the Baltic birds drastically reduced their migration distance. After the spontaneous establishment in the 1970s the number of breeding pairs rose rapidly, due to high reproductive success and low mortality rates. In recent years, breeding colonies has been spontaneously established in a range of different habitat types, including urban habitats, at more than 100 sites throughout the Baltic Sea region. From the central Baltic region the population has expanded towards the west coast of Sweden and the coastline of south-east Norway (Ree 2001).

Summer in Svalbard

Barnacle Geese begin arriving in Svalbard in the second half of May. When approaching Svalbard a most impressive mountainous landscape appears. The archipelago has, in a geological sense, a dynamic past with a large variety of volcanic and metamorphic rocks and sediments of different origin. The peaks are serrated and the mountains steep. The abundant glaciers contribute to the inspiring scenery: 60% of the area is covered by glaciers that flow out of the mountains to the sea between mountain peaks (Figure 2.2). Visitors are surprised by the barren landscape. There are no trees in Svalbard, apart from a variety of willow species whose tiny leaves lie flat on the ground. Much of the plant and animal life is restricted to the lowland areas along broad river valleys, the borders of the large fjords and, most important for geese, the flat coastal plains that extend between the steep mountains and the sea. The geese make use of vegetated areas on headlands, around the perimeters of streams and lakes. They also exploit habitats near large cliffs where abundant and highly nutritious vegetation is fertilised from seabird excrement. Seabirds that breed in Svalbard include the Fulmar *Fulmarus glacialis*, Common Guillemot *Uria aalge*, Brunnich's Guillemot *Uria lomvia*, Black Guillemot *Cepphus grylle*, Little Auk *Alle alle* and Puffin *Fratercula arctica*. Several herbivore species share Svalbard's sparse tundra resources. Interactions occur between Barnacle Geese and potential competitors, like Light-bellied Brent Goose, Pink-footed Goose and Reindeer *Rangifer tarandus platyrhynchus* (Madsen *et al.* 1989, van der Wal *et al.* 2000, Fox & Bergersen 2005).

Svalbard is located in the extreme north between 74 and 81°N, a distance of only 900km to the North Pole. However, conditions are less severe than in areas at similar latitude like, for example, Ellesmere Island, northern Canada. This is due to a branch of the Gulf Stream that touches the north-west coast of Svalbard. Still, the land has the features associated with its northern location, including continuous daylight in summer for four months, low air temperatures and little summer precipitation. Biogeographically, Svalbard is part of the high Arctic (Bliss 1981), and the sparse vegetation is typical for the polar semi-desert: a low plant cover, dominated by mosses, lichens or herbs like *Dryas*. On wetter places the tundra has elements of sedge-moss communities. Up to the 1970s the

Figure 2.2. *The western island in the Svalbard archipelago, referred to as Spitsbergen, 'land with peaked mountains', is covered with several inland glaciers. In summer geese make use of much of the coastal region, the flat area in the foreground. Photo by Jouke Prop.*

breeding distribution of Barnacle Geese was largely confined to the west coast, but with an increasing population size the geese have now colonised many of the suitable tundra areas throughout the archipelago (Mehlum 1998). With an increasing population size, the species also settled as a breeding bird from the 1990s on the islands of Bear Island and Jan Mayen, south of Spitsbergen.

We concentrated study efforts in Svalbard along the Nordenskiöldkysten (a strip of coastal area named after the Swedish geographer) located in the mid-western section of the archipelago (77°50'N), but we also kept track of observations from 17 other breeding areas during periodic catching expeditions, surveys and short-term study periods. When geese arrive most of the tundra is still covered with snow except for south-facing mountain slopes on the coast or sheltered inland valleys. It is in these places that the geese spend the first days after arrival, vigorously feeding as long as weather conditions allow, benefitting from the scarce grasses appearing from a thick moss layer (Hübner 2006). Migration tracks of birds fitted with a satellite transmitter combined with observations in the field suggest a coarse network of these pre-breeding sites (Griffin 2008, Tombre *et al.* 2008, Hübner *et al.* 2010). Moving from one site to the next, geese travel northwards, choosing the site closest to their breeding location as a final stopover place. From there they make 'inspection flights' to the future nesting location (Hübner *et al.* 2010), which may serve to assess local snow conditions or to choose a suitable nesting territory. The very first pairs settle at their nest location by the end of May (Griffin 2008), but the majority of geese arrive in the colonies during the first week of June (Chapter 11). Other pairs have a more leisurely schedule and start egg-laying as late as the end of June or early July.

Usually pairs occupy nests of previous years, but not necessarily their own. Nests consist of a depression of several centimetres in the hard soil, surrounded by a rim of pebbles and old droppings. When available, the shelter of stones or pieces of wood washed ashore may be used as a nesting place. It is a challenge for prospecting pairs to defend nest sites against competing pairs that are in a similar phase of settling (Chapter 7). Regarding the nest, not much work is needed to satisfy the needs of pairs eager to breed, and the first egg is produced soon after arrival. During the following days when the clutch is completed, females work on perfecting the nest by collecting more small stones from the immediate surroundings.

When the geese lay their eggs, breeding locations are just starting to emerge from the snow cover and, depending its thickness, different habitats on the tundra begin to appear (Figure 2.3). The first parts of the tundra that become snow-free are the crests of beach ridges and the peaks of rocky outcrops, followed by ridge slopes and finally the more extensive flat areas behind the beach wall. Moss-meadows around the perimeter of inland lakes and pools are the last vegetation zone to appear, and are a particularly important habitat for the geese (Figure 2.4). The main food species in each vegetation zone are indicated in Figure 2.3. First, seedheads produced in the previous summer (mainly the herbs Arctic Mouse-ear *Cerastium arcticum*,

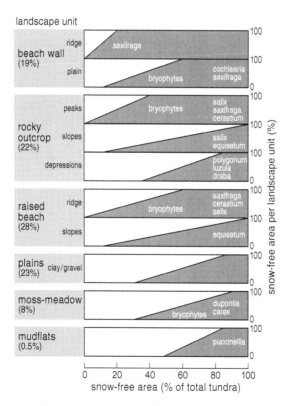

Figure 2.3. *Recession of snow from the various Svalbard habitats or landscape units in relation to snow cover of the total study area. First, the beach wall, large rocks and raised beach ridges appear from snow cover, thus giving geese access to a variety of herbs and mosses (bryophytes). Later in spring, grasses (*Dupontia*, *Puccinellia*) and sedges (*Carex*) become available in moss-meadows and mudflats. From Prop & de Vries (1993).*

Figure 2.4. *Svalbard vegetation zones important to the geese. (A) The polar semi-desert, or fjellmark, (B) wet moss-meadows, and (C) the beachwall. Fjellmark is most common on the Nordenskiöldkysten study site (> 70% of the area), and is sparsely covered with herbs, low deciduous shrubs, lichens and mosses. Moss-meadows (8% of the area) occur in depressions where drainage is impeded. The vegetation of these meadows consists of a closed moss carpet (mainly* Calliergon *spp.), with grasses (*Dupontia pelligera*) or sedges (*Carex subspathacea*). The beach wall is scattered with herbs, mainly* Saxifraga cespitosa *and* Cochlearia groenlandica. *The photographs are from the study area at Nordenskiöldkysten (Jouke Prop).*

Purple Saxifrage *Saxifraga oppositifolia* and Tufted Saxifrage *S. cespitosa*) appear, followed by Variegated Horsetail *Equisetum variegatum*, Polar Scurvygrass *Cochlearia groenlandica*, rhizomes of Alpine Bistort *Bistorta vivipara*, and buds of Polar Willow *Salix polaris*. Subsequently, grasses (tundra grass *Dupontia* spp., Creeping Alkali Grass *Puccinellia phryganodes* and *Poa* spp.) and sedges (Arctic Saltmarsh Sedge *Carex subspathacea*) appear in the moss-meadows. Mosses occur in a wide array of vegetation zones and are important food plants for much of the summer.

The thaw in the Arctic summer is highly variable. Usually, half of the tundra is snow-free by 15 June, though in recent years snowmelt is on average a week earlier than in previous decades. Once temperatures have climbed above freezing the deep snow cover may entirely disappear within two weeks. In some years, however, the snow persists well into July. The earliness of snowmelt has a large effect on the reproductive success of Barnacle Geese (Owen &

Norderhaug 1977, Prop & de Vries 1993), as it has in other Arctic-breeding birds (Remmert 1980). This is partly due to the availability of nest sites; part of the nesting locations may still be snow-covered when prospecting for nests begins. More important is the effect of the time of snowmelt on the availability and growth of food plants. The geese will only complete the incubation period when obtaining sufficient nutrients. Although they may fly as far as 2km from the colony, efficient foraging requires the availability of a variety of vegetation types.

Summer temperatures are rather constant, averaging 5°C on the coastal plains in July and August. Continuous sunlight during the Arctic summer helps mediate the constant temperatures, which rarely exceed 10°C. It commonly snows or drizzles at this time of year, but it does not amount to much. Even summed for the whole year, precipitation does not exceed 40cm, which is considerably less than, for example, in New York (118cm), Oslo (76cm), or London (75cm).

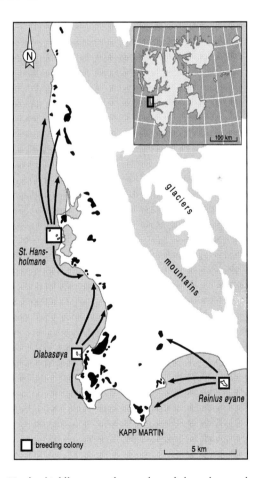

Figure 2.5. *Map of the Nordenskiöldkysten study area bounded on the west by the ocean and to the east by steep, 900m high mountains. A seawall composed of gravel and sand deposits lines the coast. Some 50 small shallow lakes are located immediately behind the seawall, and another 35 are situated more inland. The main Barnacle Goose colonies used to be on four offshore islands, but today nesting locations are more dispersed. Barnacle Goose families disperse from the islands to the freshwater ponds and lakes on the mainland tundra, as indicated by the arrows. From Prop* et al. *(1984).*

The most important habitat type for nesting in Svalbard is flat offshore islands, though the geese also breed at coastal sites (skerries or islets close to the shore), and on steep cliffs and in canyons (Prestrud *et al.* 1989). At hatch, tiny goslings must jump to the sea or rock scree below, while hungry gulls circle above their heads. When in the sea, parents and goslings must swim and navigate waves near the shore. Then goose families walk to brood-rearing habitat found on the perimeter of ponds and lakes on the mainland, usually travelling less than 10km but sometimes up to 25km (Figure 2.5). Hatching time normally coincides with the stage when food plants are most nutritious and abundant for goslings (Prop 2004). Within just six weeks goslings grow from small downies to almost fully developed geese.

Non-breeders and failed breeders use these same waterbodies to moult; there is no evidence of a moult migration. Main food plants during the moult include grasses (Tundra Grass *Dupontia pelligera*, Arctic Bluegrass *Poa arctica* and Red Fescue *Festuca rubra*), supplemented with a variety of herbs (e.g. buttercup *Ranunculus* spp., mouse-ear *Cerastium* spp.). As soon as the geese gain the power of flight they leave the moulting lakes, which are usually depleted by the intensive grazing, and they have three to six weeks left to prepare for the autumn migration. During that time they mainly use the tundra to feed on a large variety of plants, including *Equisetum variegatum*, fruits of *Saxifraga* spp., and bulbils of *Bistorta vivipara*. From the end of August onwards the geese move to lush river valleys inland and in particular, after the first snow, they use the well-fertilised slopes beneath seabird cliffs, feeding on grasses, including Foxtail *Alopecurus borealis*, *Poa arctica* and *Dupontia pelligera*.

Figure 2.6. *Bear Island (Bjørnøya) plays a key role during the autumn migration, and most of the Barnacle Geese use this isolated island. Note the steep cliffs (400m high). Goose food here consists mainly of the sparse but well-fertilised grass carpets, located adjacent to massive seabird cliffs and along streams and ponds (Owen & Gullestad 1984). The birds roost on the edge of the lakes and on rocks and stacks to escape raids by Arctic Foxes. Photo by Georg Bangjord.*

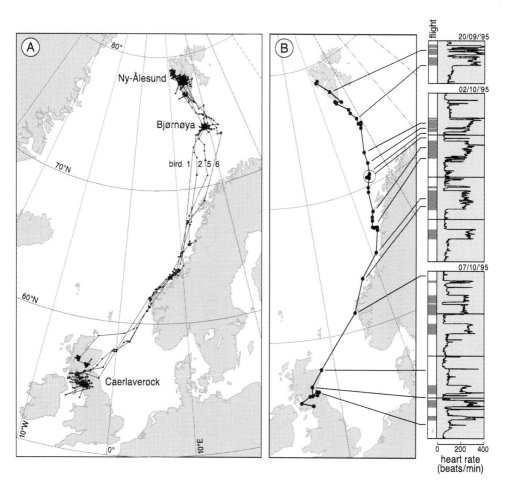

Figure 2.7. *(A) Migratory route taken by four Barnacle Geese fitted with satellite transmitters travelling from Ny-Ålesund (northern Svalbard) to southern Scotland. (B) Satellite track and heart rate data from a pair of non-breeding Barnacle Geese during their autumn migration in 1995. It is assumed that they travelled together. The bird from which the satellite track was obtained was a male and had a mass of 2,090g when the satellite transmitter was attached. The other bird was a female with a mass of 1,640g. The circled points indicate a time when the birds were resting on the sea. From Butler & Woakes (1998).*

Goose migration to the south occurs from mid-September onwards, usually following heavy snowfall and low temperatures (Prop *et al.* 1984). On departure geese gather at southern points on the main archipelago before travelling to Bear Island (74°30'N), which is 250km south of Spitsbergen (Figure 2.6). Their stopover on Bear Island may be important for enabling geese to extend pre-migration stores, a prerequisite for successful migration in some years (Owen & Gullestad 1984). Butler & Woakes' (1998) ingenious study with lightweight satellite transmitters showed that the geese continue their flight down the coast of Norway and then over to Scotland (Fig. 2.7A and B). The geese periodically stop on coastlines or in the sea, during an average non-stop flight time of 13 hours. The average flight time for the 3,100km journey from northern Svalbard to Scotland was 61 hours.

Wintering grounds

The centre of the birds' distribution on arrival each winter is the Caerlaverock area on the Scottish side of the Solway Firth (55°N). Viewing massive flocks making up the majority of the population in front of WWT Caerlaverock's observation towers in October is an annual highlight. The birds' winter range along the shores of the Solway Firth estuary extends 25km from the northernmost to southernmost haunts and 50km from the westernmost to easternmost haunts. The total area of 1,250km² is rather small compared to winter ranges for other goose populations. Furthermore, the geese use a small part of this range since foraging and roosting areas are concentrated on six areas, which receive variable amounts of use (Owen *et al.* 1987, Phillips *et al.* 2003).

The saltmarsh in this region consists of low-lying, salt-tolerant sward intersected by a network of narrow ditches that fill at high tides (sloughs). The species composition of the plant communities is influenced by local topography, livestock grazing, and wave action and ocean spray on the shore (Figure 2.8). The main goose foods in this area are grasses, including *Puccinellia*, *Agrostis* and *Festuca*. The birds also periodically dig their bills down into moist soil to extract clover stolons *Trifolium* spp. when the marsh becomes waterlogged with rain (Owen & Kerbes 1971). The size of the marsh is determined by natural and man-induced cycles of accretion and erosion along the shores of the estuary. A breakwater was erected in the 1850s at the mouth of the River Nith and a seawall (dike) along the coast

Figure 2.8. *Air photo showing WWT Caerlaverock (Eastpark Farm) pastures and ponds (lower half of photo) and the adjacent Caerlaverock National Nature Reserve, which includes saltmarsh habitat intersected by many creeks and channels running from the estuary (top of photo). Some fields are recently harvested wheat stubble, the habitat where geese will find spilled grain. Two long straight avenues extending from the farmhouse allow inconspicuous access to hides and towers from which to view the geese in the pastures.*

Location	Best time of year	Population
WWT Caerlaverock, Scotland	Oct–Apr	Svalbard
RSPB Mersehead, Scotland	Oct–Apr	Svalbard
RSPB Loch Gruinart, Isle of Islay, Scotland	Oct–Nov, Dec–Apr	Greenland
Inishkea Islands, Ireland	Oct–Nov, Dec–Apr	Greenland
Ballintemple, Ireland	Oct–Nov, Dec–Apr	Greenland
Wadden Sea coast, islands and mainland, the Netherlands	Oct–May	Russian/Baltic/North Sea
Delta area, the Netherlands	Oct–Apr	Russian/Baltic/North Sea
Wadden Sea coast, Germany	Oct–Nov, Mar–May	Russian/Baltic/North Sea
East coast of Gotland, Sweden	Apr–May	Russian/Baltic
Southern Öland, Sweden	Oct	Russian/Baltic

Table 2.2. *Locations where Barnacle Geese may be observed in the wild. These sites often include excellent facilities to enable good views of the geese and other species that visit coastal wetlands. Some also have a visitor centre and gift shop.*

separating marsh from inland pastures that were progressively drained. On the inland side of a seawall, farmers sow a more digestible food (for dairy cattle and sheep), primarily Ryegrass *Lolium perenne* and White Clover *Trifolium repens*. Pastures at WWT Caerlaverock refuge and RSPB Mersehead are managed especially to attract and hold geese and other waterbirds by grazing livestock in summer and fertilising prior to and during each goose season (Owen 1980b, Owen *et al.* 1987), a theme to which we return in the final chapter.

Birders and other countryside enthusiasts make regular pilgrimages to see the Barnacle Geese in winter on the Solway Firth at Caerlaverock and Mersehead, Scotland. The WWT Centre at Caerlaverock has four observation towers and several smaller hides (blinds) that are approachable from behind tall embankments (Figure 2.8). Table 2.2 lists other sites where Barnacle Geese may be seen in large foraging flocks, including the Isle of Islay in the Scottish Hebrides, the Dutch and German Wadden Sea coast, and the Frisian Islands in the Netherlands.

Spring staging habitats

On their return journey to Svalbard from the wintering areas in Britain the geese stop to refuel on the coast of Norway. There are two main regions used by the geese. One is called Helgeland, located just south of the Arctic Circle (65°45'N). The other region is Vesterålen, 350km to the north and well above the Arctic Circle. Helgeland accommodated most of the birds for a long time but its relative importance has decreased in the 21st century. It consists of about 10 archipelagos each with hundreds of small, flat, maritime islands. Altogether, the geese may utilise vegetation on more than half of the 10,000 islands during their stay in April/May. However, the quality of goose food on the islands varies enormously due to topography, oceanic climate and, of course, man's use of the islands. The islands furthest from the mainland used to be the stronghold for the geese (Gullestad *et al.* 1984), but since the 1980s they have colonised larger islands to the north and east closer to the mainland. Sections of these larger islands are cultivated and managed as pasture and meadow, while the coastlines consist of salt-tolerant vegetation. Out in the open ocean the islands become smaller and flatter; usually they are less than 10m in height. These windswept islands look barren and have no trees. The higher and driest parts are covered with a nutrient-poor heath vegetation with shrubs like *Juniperus* and *Cassiope*. Plant communities that are important to geese include: (i) herb-meadows with herbs like *Alchemilla* and *Filipendula,* and some grasses, (ii) marshes dominated by *Festuca rubra*, and (iii) marshes dominated by *Puccinellia maritima*.

The remote outer islands used to be inhabited by hardy farmer/fisherman families who shaped the land for their sheep and dairy cows. These people changed the highest parts of some of the islands into small hay-meadows and pastures composed of high densities of a variety of grasses (*Poa, Agrostis, Anthoxanthum*). The soil of the islands where they lived (home islands) was built up by the traditional practice of hauling seaweed onto the island, where it was composted and cultivated into pasture. Small sheep flocks were ferried to the outer islands to graze in spring and summer. Hay for feeding livestock in winter was cut by hand and dried on the home island. Manure from the livestock was applied to hay-meadows. Clearly, the agricultural value of these hay-meadows and small pastures was in no way comparable to the intensively managed fields of the larger islands close to the mainland where machinery and tractors were employed. However, the management of home islands produced a shorter and better sward than the vegetation on outer islands, which in spring were much dominated by thick tussocks of dead material. In the 1970–80s the majority of the inhabitants of the remote islands moved to the larger islands or to the mainland, and many of the archipelagos became abandoned, though not by the geese, as we will explain in later chapters.

People of the remote islands afforded a tolerant attitude toward the geese that were 'welcomed as a sign of the onset of spring'. Geese were not chased from the pastures, and as a consequence they foraged among the crofters' houses and barns soon after arrival (Figure 2.9). This behaviour is contrasted to that of other goose species that are harvested in the region at other times of the year. However, in the case of the Norwegian fishermen, some self-interest was involved in not disturbing waterfowl. Part of their economy was based on a long-held tradition of collecting and processing the down of Common Eiders *Somateria mollissima* (Wold 1985, Hatten & Norderhaug 2001). For this purpose, home islands were littered with Eider huts that were lined with a bed of clean, dried seaweed where female Eiders established their nests. Each morning as the sun rose at about 02:00 hrs till 09:00 hrs, Eiders came ashore from the surrounding water to prospect for nests. During the day Eider eggs and down were harvested.

The northerly Vesterålen attracted Barnacle Geese from the end of the 1990s onwards (Shimmings & Isaksen 2013, Tombre *et al.* 2005, 2013a). Traditionally, Vesterålen had been an important stopover site for Pink-footed Geese (Madsen 2001). In this region the geese feed on the narrow strips of land between the coast and higher grounds that are used by local farmers as pastures and hay fields. Grass crops grown on the agricultural fields are mainly composed of Timothy Grass *Phleum pratense* and saltmarshes lining the coast are dominated by Red Fescue *Festuca rubra* and saltgrass *Puccinellia* spp. (Tombre *et al.* 2005). Due to a strong competition for the spring growth of grass between farmers and geese, geese were chased intensively by the agricultural community. From 2006 management actions were

Figure 2.9. *Geese visit the 'improved' sward on the Home Islands during the early morning (daylight persists for 22 hrs in late May). Note the geese between the buildings. This picture shows the Lånan archipelago consisting of about 300 small, flat islands where the geese find newly emerging, sea-swept vegetation. The 'Eider houses' are part of the outer islanders' traditional lifestyle. The upturned boats contain several nests, smaller constructions only one (lower panel). Photo by Jouke Prop.*

implemented to alleviate conflicts between farmers and Barnacle and Pink-footed Geese (Tombre *et al.* 2013a,b; Chapter 15).

From the mid-1990s onwards, the Svalbard population adopted a new strategy (Griffin 2008, 2012). Rather than using one of the Norwegian staging regions as a stopover site, an increasing proportion of the population remained on the Solway Firth much later in the spring. By employing satellite transmitters, Griffin (2008, unpubl. obs.) showed that some of these individuals migrated to Svalbard almost directly, spending only a few days in the

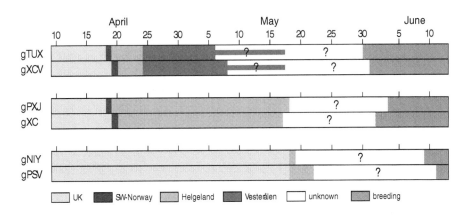

Figure 2.10. *Patterns of spring migration in the Svalbard Barnacle Goose population, as indicated by the geographical location between 10 April and 10 June. Three different migration strategies were distinguished with geese spending most of the spring staging period (end of April through mid-May) in Vesterålen (upper panel), Helgeland (middle), or the wintering range in the UK (lower). Data were obtained from 'geolocators' attached to one of the leg-rings. Data from M. J. J. E. Loonen and T. Oudman (unpubl. obs.).*

Norwegian staging sites. Other geese, leaving the Solway at the normal time, spent 18 days in Helgeland and a further seven days, including a two-day stop at Vesterålen, en route before touching down on southern-facing slopes of Spitsbergen. Maarten Loonen and Thomas Oudman benefited from new technologies in tracking devices by fitting geese with 'Global Location Sensing loggers', or geolocators. These are small (9g) and relatively cheap devices attached to the leg-rings. The loggers measure and store information on light intensity, which provides a way to calculate the global position for each day between deployment of the device and moment of retrieval (memory and batteries allow data storage for several years). The logger data supported the three distinct migration schedules: the 'Norwegian' birds departed from the Solway usually well before 5 May to spend much of May in Helgeland or Vesterålen, whereas the 'Solway' birds spent another couple of weeks in the UK (Figure 2.10; Oudman 2009). During the continuous daylight of the Arctic summer, and during the equinox, the loggers do not provide accurate data. As a consequence, it becomes hard to calculate locations of geese staging in Vesterålen after 6 May (as indicated by the dashed line with question mark in Figure 2.10). The timing of incubation, though, can be reconstructed nicely thanks to the dark periods when the female is sitting on eggs (and covering the data logger attached to her leg-ring).

Spring and summer on Gotland

The first spring migrants arrive at spring staging sites on Gotland in the second part of March and peak numbers of spring-staging birds can be observed in the first half of May, coinciding with the emergence of initial 'spring' growth of their main food plants. While the Arctic Russian geese are still present in the staging areas, the Baltic Barnacle Geese already start nest building and laying eggs at the end of April (Figure 2.11). Those that depart towards the Russian Arctic do so in the second part of May.

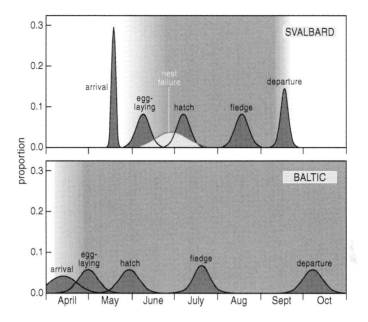

Figure 2.11. *Breeding timetable of Svalbard and Baltic populations compared. Arctic birds have a tight schedule with the tundra starting to clear from snow (indicated by the shading) only after incubation has begun. In autumn, snow forces the geese to migrate south. Baltic birds arrive one month earlier on the breeding grounds, coinciding with the disappearance of most of the snow. Foraging conditions in the Baltic remain favourable through a large part of the autumn and geese stay well into October. The long season for the Baltic birds may be one of the reasons for the lower synchrony in the breeding cycle. Goslings in the Baltic grow at a slower rate, at only 70% of that of the Arctic birds. As a result, they need several weeks more to fledge. Compiled from Loonen* et al. *(1997a) and Prop (2004).*

The nesting islands along the coast of Gotland are rather flat and vary in size from less than 1ha to more than 1,000ha. The larger nesting islands are covered with grasses (e.g. *Festuca, Agrostis*) and herbs that are grazed in summer by sheep or cattle (Figure 2.12). Trees, small forest patches and bushes can be found on some of the nesting islands, which also host a large number of other breeding birds like gulls, terns, ducks, waders and cormorants.

Compared to the Arctic, the summer climate in the Baltic is luxurious. Gotland is situated within the north European mixed forest zone (Moen 1999) and the average temperature in July is close to 20°C. However, at the end of April when the Baltic-breeding Barnacle Geese start nest building the weather at coastal breeding sites can be quite harsh and not very different from the weather nest-building Arctic-breeding geese are exposed to about one month later. The Gotland colonies are usually snow-free by mid-April but the temperature during the night is often below, or close to, 0°C. Periods with strong cold wind, heavy rain, or even snowfall are common during the incubation period.

Unlike their Arctic counterparts, breeding geese on Gotland feed on livestock-grazed shore meadows and agricultural land during spring, summer and autumn. During incubation the breeding birds mainly feed close to their nests within the colonies. During the brood-rearing period from the end of May to the end of July, most family groups leave nesting islands to feed on larger shore meadows along the coast. Non-breeders and failed breeders feed both

Figure 2.12. *Barnacle Goose colonies on Gotland are usually situated on flat islands with low vegetation of grasses, herbs and bushes. This is Storholmen, the largest island of the island group Laus holmar. The island group hosts the oldest Barnacle Goose colony in the Baltic region. Gotland is the background. Photo by Gunnar Britse.*

on shore meadows and on cultivated agricultural land. During wing moult in the middle of July, when non-breeders and failed breeders as well as brood-rearing adults are flightless, the geese are confined to a smaller number of shore meadows very close the sea where they have the possibility to retreat from disturbances and mammalian predators. From the end of July, when goslings fledge and adults regain flight capacity, until September, the geese on Gotland often leave coastal habitats to feed on inland agricultural land. Baltic Barnacle Geese as well as the returning geese of the Arctic Russian population migrate to the wintering grounds in the Netherlands and Germany in October and November.

Goose predators

Any description of study sites must include a list of predators, one of the main influences on goose behaviour and distribution. Geese are regularly on the lookout for potential predators. One of the most obvious features of a goose flock or colony is that a certain proportion of its members stand with head and neck erect in a vigilant, scanning posture (Drent & Swierstra 1977, Inglis & Lazarus 1981, Forslund 1993).

In Svalbard, Arctic (or Polar) Foxes *Alopex lagopus* are the main predators of geese, taking eggs, goslings and even adults when given the opportunity (Figure 2.13A). They are ubiquitous and sometimes numerous, particularly near seabird colonies. When Polar Bears swim to island colonies they can destroy a vast number of nests (Madsen *et al.* 1989, Drent

& Prop 2008, Prop *et al.* 2013, Figure 2.13B), or chase moulting birds (Stempniewicz 2006). Chapter 14 describes the increasing impact of 'ice bears' in the breeding season. Other predators of eggs and young goslings (less than two weeks of age) include Glaucous Gulls *Larus hyperboreus* and Great Black-backed Gulls *Larus marinus*, Arctic Skuas *Stercorarius parasiticus* and Great Skuas *Stercorarius skua*. Outside the breeding season predation events are rare but the geese seem well tuned to the possibility of attacks (Patterson 1995, 1998). On the Solway Firth we have seen several geese that were knocked out of the sky by

Figure 2.13. *(A) The Arctic Fox is one of the main predators of Barnacle Geese in Svalbard. In years when foxes reach nesting islands via ice bridges, geese may delay or forego nesting altogether. Foxes living near goose colonies on cliff faces patrol the scree for injured goslings during nest exodus (Christine Hübner). (B) Polar Bears have become important predators of goose eggs. In this picture a female bear plunders the nests in our main study colony. Changes in sea ice distribution bring more Polar Bears to the west coast of Svalbard, where most goose colonies are found (Jouke Prop & Jim de Fouw). (C) The main predators of eggs and goslings in Gotland are large gulls, in this case a Lesser Black-backed Gull (Kjell Larsson).*

immature Peregrine Falcons testing their abilities. The only other predators on the wintering grounds are Red Foxes *Vulpes vulpes* that patrol the mudflats at night and the marsh by day for unsuspecting geese. In Helgeland in spring, the geese are also concerned with aerial predators, including Peregrines, Gyrfalcons *Falco rusticolus*, and White-tailed Eagles, which cause geese to panic and take flight several times per day.

In the Baltic, Herring Gulls *Larus argentatus,* Greater Black-backed Gulls and Lesser Black-backed Gulls *Larus fuscus* are the most important aerial predators on eggs and goslings (Figure 2.13C). Predation by gulls on goslings is more common than predation on eggs. In some years and colonies, gulls remove more than 90% of the hatched goslings before fledging. Red Foxes, and in Finland and Estonia also Raccoon Dogs *Nyctereutes procyonoides,* occasionally reach nesting islands, destroying a large number of nests and killing many breeding birds. The White-tailed Eagle, a species that was almost extinct in the Baltic region in the 1970s, is since the 2000s a common coastal species in the central Baltic region. White-tailed Eagles, including specialised individuals, regularly attack adult and juvenile Barnacle Geese and other waterfowl in the breeding colonies. Barnacle Goose remains (and leg-rings) have also been found in nests of Golden Eagles *Aquila chrysaetos.* Goshawks *Accipiter gentilis* and Peregrine Falcons may occasionally attack adult geese. In subsequent chapters, we will describe some of the consequences of living with the risk of predation in goose habitats.

Barnacle Geese in perspective

The 15 goose species of the world are subdivided into two genera, *Branta* (black geese) and *Anser* (grey geese, including *Chen*). Molecular phylogenetic analyses of *Branta* geese indicate that the Barnacle Goose and the small races of Canada Geese share a common ancestor (Paxinos *et al.* 2002). Other more distantly related geese in the *Branta* group include the Hawaiian Goose *Branta sandvicensis*, large races of Canada Geese, all races of Brent Geese *Branta bernicla*, and Red-breasted Goose *Branta ruficollis*. Apart from the darker colours in *Branta*, a striking difference between genera has to do with their bills (Owen 1980a). *Branta* have soft bills without serrations, whereas *Anser* bills are generally harder and serrated. This difference enables exploitation of different foods: most *Branta* specialise in grazing and seed stripping, whereas most *Anser* are also capable of biting into harder foods (e.g. turnips and carrots) and grubbing below ground for tubers.

Branta- and *Anser*-like geese appeared in the fossil record about 4.5 million years ago, but modern-day Barnacle Geese split from their small Canada Goose ancestor by moving to the Palaearctic less than 0.5 million years ago (Paxinos *et al.* 2002, R. C. Fleischer, pers. comm.). Given that there have been several long-lasting glacial periods since then, the Barnacle Goose range must have shifted with the coming and going of suitable foods and breeding locations. The geese would have been forced to retreat southward as the glaciers advanced (see also Ruokonen 2001). In the last 150,000 years alone, there have been three periods of ice expansion (Elverhøi *et al.* 1998). Figure 2.14 shows a hypothetical migratory corridor and range for Barnacle Geese during the last ice age. The entire current-day distribution of the species was engulfed by huge ice-domes in this late Weichselian period (Anderson 1981, Denton & Hughes 1981, Elverhøi *et al.* 1998, Landvik *et al.* 1998). Current-day breeding areas in Greenland and Svalbard were covered by up to 500m of ice, and those in Russia by

Figure 2.14. *Extent of ice coverage during the most recent glacial period and a hypothetical southern distribution of Barnacle Geese during the Pleistocene. As recently as 10,000–40,000 years ago several large ice-domes covered much of the region from England through Russia and from Iceland and Greenland through Canada. The Barnacle Goose has been uncovered at 18 archaeological sites, 14 of which are indicated on this map. Sites were located in England (n = 1), Ireland (1), Jersey (1), the Netherlands (1), Germany (1), France (2), Spain (3), Italy (4) and Malta (4) (Tyrberg 1998). Three of the fossils from Italy were dated back to a previous interglacial period more than 127,000 years ago. Barnacle Geese may have bred on the steep coastal cliffs and islands in Scotland, England, Wales and Ireland and wintered in the wetlands around the Mediterranean. The arrows indicate possible migration routes.*

as much as 2,000m of ice. The ice began retreating from current-day wintering and staging areas about 12,000 years ago. Breeding areas in Spitsbergen (Svalbard) became available about 10,000 years ago, followed by areas in east Greenland and Russia about 2,000 years later.

The first evidence of use of the current range for the Svalbard Barnacle Goose population comes from a set of bones dating to 3,470 years ago found in a cave on the Norwegian coast (Vega, 65°N) in Helgeland, Norway (Lie 1989). During periods of glacial change, selection would have favoured genes for pioneering to new areas, causing a shift in the migratory range as southern areas became too warm. One of the main tenants that we will be developing in this book has to do with this resourceful nature of wild geese. They seem to be very adaptable to new foraging and breeding opportunities. Even over the course of our study covering just a few decades we witnessed an extreme shift in the type of food geese eat, from natural saltmarsh to agricultural plants, and major changes in migratory behaviour to allow the establishment of new colonies outside the birds' traditional Arctic range.

Summary

Our study focused on two of the five Barnacle Goose populations, one that bred in Svalbard, Norway, and another in the Baltic Sea region. Founders of the Svalbard population may have mixed with the Greenland population in winter at the turn of the 20th century. The early population experienced heavy exploitation during the breeding and non-breeding seasons from egg collecting and hunting. Our Arctic study population bred in Svalbard just 900km from the North Pole. During autumn migration the geese stopped on Bear Island and the west coast of Norway. They wintered on the border of England and Scotland on the estuarine and agricultural habitats of the Solway Firth and on their way back north in spring stopped again on the west coast of Norway to refuel, primarily in a region of 10,000 sea-swept islands called Helgeland. Our second study was conducted in an area of the Baltic Sea region that is used by migrating geese travelling to Arctic Russia. Some Russian birds began to nest on islands off the coast of Gotland, Sweden in 1971. By 'short-stopping' to breed at Baltic 'staging areas' in this way, the birds reduced their migration by 2,000km.

The main predators of Barnacle Geese in summer include Arctic Foxes, Polar Bears and large gulls in Svalbard and large gulls, Red Foxes and White-tailed Eagles in the Baltic. At other times of the year, large birds of prey and foxes continue to influence the birds' behaviour and distribution. The migratory pathways of wild geese must have changed several times to accommodate the position of glaciers during the ice ages. Just 10,000 years ago huge ice-domes covered current-day Barnacle Goose habitats, shaping the availability of vegetation on which geese depend. Even on the timescale of decades, the size and distribution of our two study populations has been dynamic. The concept that geese are well able to track their food plants over time and space is a recurring theme in the book.

CHAPTER 3

Research methods

In this chapter we describe the methods used to monitor and analyse characteristics of individual geese, flocks and populations. We also describe methods for tracking growth and depletion of goose food plants. For some of the more involved methods requiring formulas for calculations we provide separate sections (Boxes 3.1–3.3) that can be referred to when more detail is required. We begin by acknowledging that the choice of techniques to study wildlife is often a compromise between obtaining quality data to test particular research questions and the costs of employing the techniques; costs in terms of the need to avoid disturbance to study animals and their habitat, efforts by the observers, and of course the research budget.

Population assessments

Population assessments for the Svalbard Barnacle Goose population were made each winter when the birds were congregated at relatively few locations on the Solway Firth. The Wildfowl and Wetlands Trust (WWT) and associates began monitoring the size of this population in the winter of 1959. Since 1970, in addition to flock counts, brood sizes and the percentage of juveniles were also assessed each winter; the less dark and defined juvenile Barnacle Goose plumage is readily distinguishable from that of the adults. The count effort was increased as the population and wintering distribution expanded (Owen *et al.* 1987, Black *et al.* 1999). Coordinated censuses were accomplished by around 12 expert counters at the main haunts and flight paths in the 1,250km² range. Sometimes population estimates included flocks reported by experienced counters along the birds' migratory route in north-eastern Scotland en route to the Solway Firth. As numbers increased still further a more sophisticated set of criteria was established to confirm annual estimates. Due to count variability and the possibility of double-counting, final count totals for the population were derived by averaging counts from multiple coordinated surveys that were within 10% of the maximum count during the winter (rounded up to the nearest 100; Griffin 2009).

In Norway, Barnacle Geese were monitored at staging areas during their stay in April and May. Counts in the early part of the study were made from the highest vantage points on the 'outer islands' in Helgeland, usually roof tops of the crofters' homes (Gullestad *et al.* 1984). When geese began to use agricultural fields closer to the mainland, we travelled by bike or car to conduct counts from strategic high points on these larger islands. Such surveys have continued for the last 20 years and coverage has increased throughout the Helgeland region (Shimmings & Isaksen 2013) and more recently the Vesterålen region (Tombre *et al.* 2013a). A smaller proportion of the entire population has been counted in Norway, in part because there are more geese, but also because the birds make use of many more of the coastal communities and islands in western Norway. A larger proportion of the population was located in three years with more extensive coverage by boat or aerial surveys.

The birds' distribution in Svalbard (May through September) was also scattered among numerous islands and coastal areas. Travel among islands was achieved with large boats capable of manoeuvring through pack ice conditions (Prestrud & Børset 1984, Prestrud *et al.* 1989, Bustnes *et al.* 1995). Smaller boats were used to land and count nesting pairs from two or three locations around the perimeter of colony islands. Landings were chosen to minimise disturbance. Since the geese and other wildlife in the area were easily scared, air surveys were rarely employed. Since 2009, records of nesting and moulting geese in Svalbard have been centrally collected from all stakeholders, including expedition scientists, agency managers and bird watchers (Tombre *et al.* 2012).

In contrast to the Svalbard population, which is separated from other Barnacle Goose populations all year round, the Baltic population mixes with the Arctic Russian and North Sea populations in winter and spring, making specific winter counts of Baltic breeding birds unachievable. Instead, we counted the total number of incubating females and nests in colonies on Gotland during visits in May. Counts of nesting birds in the oldest and largest colony were performed from strategically located towers and hides. The number of nesting pairs in other parts of the Baltic region was also regularly counted, and other researchers or amateur ornithologists have published these results. Counts of non-breeders and failed breeders during summer have been impractical since these categories of birds were scattered on a large number of sites along the coast and on inland agricultural land. Therefore, we relied on indirect methods to obtain estimates of the total number of birds belonging to the Baltic population. This was achieved by combining counts of nesting individuals and fledged juveniles produced in different years, adjusting for estimated survival rates and proportions of non-breeders in different age groups (Larsson & van der Jeugd 1998). For the Baltic population, estimates of the number of breeding pairs were usually directly derived from counts of nests and incubating females with a high degree of reliability. Estimates of the total number of individuals that belong to the Baltic Barnacle Goose population had more uncertainty.

Goose catching during the moult

Catching geese, as with any wild animal, is an art as well as a science. The key is to know how to get close enough to them before they run or fly away and then once captured to ensure that each animal is restrained properly, keeping the possibility of injury and stress to a minimum.

Waterfowl replace flight feathers (primaries and secondaries) on their wings during the summer moult. Once the old feathers fall out geese are grounded for about four weeks (Owen & Ogilvie 1979). During that time adults and their flightless goslings live on the perimeter of tundra lakes or along the coast and immediately retreat to the water on seeing a fox or human.

Our catching efforts on Svalbard peaked at the end of July when most parents and goslings were still flightless; most geese were flying again by mid-August. The roundup procedure began by carefully approaching a brood-rearing area. If the geese saw or heard people coming, even if still kilometres away, they immediately ran in the opposite direction. It is possible, therefore, for an unsuspecting person to hike along a stretch of coastline without seeing a single goose because they 'disappear' well ahead of the walker. By becoming familiar with the terrain it was possible to surround a brood-rearing area without being seen by the ever-vigilant geese. Once in place our catch team would 'appear' in unison and the geese would retreat to the middle of the nearest lake rather than sprint across the tundra or into the sea. Three or four people standing around the perimeter of the lake 'held' the birds while another three or four workers erected catching nets in the shape of a 'V'. Ideally, nets were placed at the narrowest point between adjacent waterbodies where geese normally exit. The 'arms' of the net were stretched 50–150m along both banks and a corral was positioned at the base of the 'V'. Nets were about one metre in height and made of sturdy fabric cord. They were held in place with fence posts pounded into the permafrost just 20cm below the surface of the tundra vegetation. Using an inflatable dinghy or kayak, geese were gently

Figure 3.1. *A catch of over 350 Barnacle Geese on Gotland at the Baltic study area. Note the double corral and associated holding pens into which geese were herded. The geese wait in a calm state once inside the canvas holding pens. Photo by Kjell Larsson.*

herded down the funnel into the corral. On the perimeter of the lake, field workers helped by wading into the very cold water. Many of the lakes were less than a metre in depth due to the permafrost layer below the water surface; a potentially dangerous situation when the ice breaks! On some occasions, we used motorboats on the ocean, pushing geese into nets erected on the shore. Once the geese were in the corral, they were transferred into smaller tents that provide warmth and shelter.

On Gotland catch efforts peaked in the middle of July because most geese were able to fly again somewhat earlier, at the end of July. A similar catch procedure was employed except the birds were gently pushed off the sea with small boats or kayaks into the 'V' trap on dry land. Two separate corrals were erected at the base of the 'V', constructed of soft canvas material, which helped reduce potential injury during large catches (Figure 3.1).

Catching by rocket and cannon nets

In the winter geese were captured by shooting rocket-propelled nets over the edge of an unsuspecting flock. This was achieved by folding the net on top of itself like an accordion and attaching a cord to the leading edge of the net. Cords were attached to a series of rockets (or projectiles shot from cannons) positioned at an appropriate angle to carry nets forward over the flock. Once the rockets were 'dug in' to the ground and the nets were camouflaged with vegetation from the surrounding habitat, a long wire was run from the rockets to a hiding place where an observer was positioned with the ignition box. Some decoy geese were positioned and a sack or two of wheat or corn strategically scattered to lure the geese into the catch area. The waiting began once the nets were set. Nets were usually set at night in the last field the geese used before departing to the roost. We hoped that at least some of the birds would return to this field the next morning to discover the grain, but this was not always the case, especially if the first birds to circle overhead suspected something, gave alarm calls and returned to the roost. So, it often took a long time for a flock to land in the right field and even more time until they marched across the field in the correct direction; goose flocks travel directly into the wind so the nets had to be placed according to prevailing wind conditions. Apart from the wind, many things could go wrong even after the birds were in the field. For example, a predator or an innocent human could flush the birds. Even when all goes well and the geese land in the right place, catch attempts have been foiled by one or two that settled in to take a nap on top of the camouflaged nets. When the geese were finally positioned in the catch zone, nets were launched and workers ran out, extracted the birds and placed them in large sacks or in tents that were erected as holding pens.

Sex, age and body measurements

Once the geese were captured and safely secured, one bird at a time was removed from the holding area and carried through a processing line where leg-rings were fitted, sex and age determined and body measurements recorded.

Geese are sexed by cloacal examination, a procedure that takes care and practise (Hanson 1967, Owen 1980a). Pressing either side of the cloacal opening exposes reproductive organs.

A fully developed male sex organ, the bursa, is difficult to miss: it is about two-thirds of the width of a pencil and is 2–4cm long when uncoiled, with some having fleshy barbs and ridges on the side. Cloacae of mature females are noticeably bigger around with stretch marks produced from passing eggs through the fleshy opening. We also checked each breast region for signs of a brood patch; either exposed skin or a patch of recently replaced feathers. This was a useful characteristic to confirm that the goose was properly sexed since only female geese incubate eggs.

Geese were aged by examining plumage characteristics. In summer catches on Svalbard we could distinguish yearlings from adults of two years and older by feather colouration and shape on the wings and tail (Hanson 1967, Ogilvie 1978, Owen 1980a). For example, whereas adult secondary coverts have dark black and grey colouration, yearlings may still be moulting brownish juvenile feathers that are grown during their first year. Birds aged as adults were assigned a minimum age by assuming they hatched at least two summers prior to ringing, whereas yearlings hatched the previous summer.

By contrast, in summer catches on Gotland it was not possible to reliably distinguish yearlings from adults of two years and older. A large proportion of the birds that were known from ringing to be yearlings did not have detectable brownish juvenile feathers, apparently because of an advanced moult schedule in this population. Hence, on Gotland birds aged as adults were assigned a minimum age of one year.

In both populations goslings which hatched in the most recent summer were easily recognised in catches by the presence of downy feathers and juvenile plumage colour. By observing brood-rearing families between hatching and fledging, the age of goslings was classified according to their size and feather development, which allowed us to estimate their hatch dates (Yocom & Harris 1966, Owen 1980a, Larsson & Forslund 1991).

Dzubin & Cooch (1992) provide excellent diagrams for several measurements that can be used to describe body size in wild geese. We found that total skull length (head + bill) and tarsus bone length (measured to the nearest 0.1mm with callipers) were the two most manageable and repeatable measurements to obtain, providing a meaningful index of

Figure 3.2. *Measuring skull length with callipers. Photo (by Sharmila Choudhury) shows Jeff Black with Maarten Loonen, our colleague of many years (holding the bird).*

body size (Figure 3.2). On occasion we also measured webs, toes, culmen, keel, metacarpal bones and wing cord. For flightless birds we measured the amount of new feather growth by inserting a thin flexible ruler between the 9th and 10th primary and measuring the length of the 9th, which usually grows to become the longest feather. The progress of feather development (i.e. the moult stage) could then be assessed on different catch dates through the summer (Owen & Ogilvie 1979, Larsson 1996). We averaged skeletal measurements made for individuals in different years to minimise effects of potential measurement error. Repeatability of structural body measures collected on the same individuals in different years was high (Choudhury *et al.* 1996). Skeletal growth is negligible in Barnacle Geese after the first year (Owen & Ogilvie 1979).

We also recorded body mass using a spring balance or a more accurate electronic balance. Development of body mass in goslings is known to be more sensitive to variations in early growth conditions than skeletal characters (Cooch *et al.* 1991b, 1996, Larsson & Forslund 1991). Weights of adults during moult vary to a large extent with the birds' breeding status (e.g. non-breeders, failed breeders, successful breeders) and moulting stage so we were cautious about how body mass was used in analyses (Owen & Ogilvie 1979, Choudhury *et al.* 1992).

Reading leg-rings

The key to success in this study was our ability to resight individually marked geese from a distance throughout the year and in many cases over the course of the birds' lives. This was achieved by fitting each captured bird with a coloured plastic ring on one leg and a metal ring on the other (or plastic rings on both legs and metal ring on one leg for Gotland birds). Metal rings had the address for one of the national ringing programs in Britain, Norway or Sweden, depending on the catch location. Plastic ring loss was estimated in the Svalbard population at a rate of only 0.35% per annum based on 18 missing rings on 'metal-only' birds that were recaptured compared to those that were still intact on recapture (Rees *et al.* 1990). Plastic rings were engraved with one to three alphanumerics. Because the plastic is composed of different coloured layers the engraving exposes a contrasting colour for the letters and/or numbers (Ogilvie 1972); for goose rings the top coloured plastic layer is 0.5mm and the bottom layer is 1mm thick. Our favourites offering the easiest read were white or yellow rings with black lettering (Figure 3.3).

During the study, 8,500 leg-rings were fitted to Svalbard birds and 407,500 resightings of these were made in the field. Each year we resighted over 95% of the marked birds still alive and recorded each one an average of nine times. We were able to achieve this because the birds frequented habitats bordered by observation towers and hides (blinds), offering an elevated position from which to look down through the grass at their legs (Figure 3.4). The two mile-long embankments built at the WWT Caerlaverock reserve enabled inconspicuous access to the pastures and saltmarsh without scaring foraging geese (see also Figure 2.8). Once in position, we used spotting scopes to read rings up to 300m away. With a little practise ring-readers became adept at watching goose legs and waiting for each of the alphanumeric codes to come in view. 'Ring-reading' was addictive for many of us – working out pair and family status for each ringed bird. There is huge entertainment value in watching the varied

Figure 3.3. *Barnacle Goose in Svalbard with leg-ring. The individual code can be read from a large distance by telescope. This bird, CIC, was seen 79 times during his five-year life, terminated during the wing moult. We found the remains of the body, and the ring, at an Arctic Fox den (bottom picture). Photo by Jouke Prop.*

Figure 3.4. *Towers and blinds at the WWT Caerlaverock refuge are ideal for observers who keep track of individually marked geese during the winter. Tall earth-filled dikes on either side of long avenues allowed us to approach flocks without being seen. Note the embankment with tall hedgerow at the top of this photo, leading to an observation tower (top left) with view to pastures. Photo by Jouke Prop.*

and amusing goose antics, including haphazard running, rolling and splashing during bath time followed by meticulous preening sessions and one-legged yoga stretches. There are also the more boisterous bouts of yelling and neck waving with family members (i.e. Triumph Ceremonies) and the humorous occurrences of 'goosing' unsuspecting neighbours from behind. Many amateur ornithologists have discovered this 'pastime' and have contributed observations of marked Barnacle Geese from the wintering grounds.

A database of more than 100,000 records was made of Gotland-ringed birds resighted on the wintering grounds in the Netherlands and Germany. Resightings of marked geese from both studies were used to estimate survival rates of different categories of individuals (Owen 1982, Bell *et al.* 1993, Larsson *et al.* 1998, van der Jeugd & Larsson 1998, Prop *et al.* 2004).

In addition to resighting a bird's ring-code and colour we attempted to identify the mate and count the number of goslings in the family unit, giving a measure of the pair's nesting status or brood size (see below). We also routinely recorded the date, flock size, position within the flock or colony, a code for the particular field, farm, colony or marsh location, and an index of the bird's body condition referred to as the abdominal profile index.

Abdominal profile index (API)

Geese deposit the majority of their fat stores in the abdomen, which in Barnacle Geese is conspicuously outlined by the white plumage between their legs, stretching to the base of the tail. The abdominal profile index is a visual assessment of the 'fatness' of this abdominal region, a method originally described by Owen (1981b). We use an index based on a scale of 0–7 where 0 is concave, 1 is a straight line from the base of the tail to the top of the legs, and 2–7 are increments of bulging or sagging toward the ground (Figure 3.5). Experienced observers can score the fatness of the birds in the field in an accurate and reliable way by regularly referring to the drawing of the index shapes that is kept in each field notebook. We

Figure 3.5. *Abdominal profile index in Barnacle Geese indicating the degree of abdominal bulging between the legs and the cloaca (near the tail) where fat is deposited. The index begins with a concave shape (rank 0) followed by a straight line between the legs and cloaca (rank 1). Fatter geese, indicated by a more rounded abdomen, are ranked with values between 2 and 7 (from left to right). Observers wait until the bird's body is parallel to the ground (e.g. head-down feeding posture) before making the assessment. The bird in this drawing is portrayed in active foraging posture when approaching the next patch of vegetation.*

Figure 3.6. *Dummy geese made of plywood are useful for consistent abdominal profile index assessments among observers and years. The picture shows the geese in their Norwegian spring habitat. Photo by Jouke Prop.*

also placed a set of plywood cut-out goose shapes with different fatness profiles in the field to help us maintain consistent assessments within and among years (Figure 3.6).

Measurements similar to the abdominal profile index are used in a variety of wildlife studies. In geese, the abdominal profile index is linearly related to actual fat reserves and body mass measurements (Féret *et al.* 2005, Madsen & Klaassen 2006). Zillich & Black (2002) report on a study of Hawaiian Geese that roamed the grounds at the WWT Slimbridge centre. Throughout the birds' annual cycle they quantified the abdominal fatness of 100 individuals that were coaxed onto a portable scale with pieces of bread. They found that the abdominal profile index was well correlated with body mass in both sexes at all times of year, except for females during the egg-laying period when profiles bulged even more. Average fat deposition rates were assessed for different periods of time, particular classes of birds (e.g. males, females), and in some cases for marked individuals that were recorded on a daily basis during the spring staging period (Prop *et al.* 2003).

Reproductive success: breeding area

To achieve our goal of identifying which individual goose types are successful we must quantify their ability at producing surviving offspring. We made these assessments during two phases. On the breeding grounds, we found that many birds attempt to reproduce while others forego egg-laying altogether. Of those that do attempt to breed, many fail along the way. Assessments of brood sizes on the wintering grounds, developed in the next section, allow us to determine which individuals successfully jumped the breeding season hurdles, thus augmenting the wintering population.

Figure 3.7. *Observation towers give a good view over breeding colonies. This hide, made of driftwood, enabled us to watch a large colony in Svalbard without causing any disturbance. Photo by Jouke Prop.*

By continuously observing breeding colonies from observation towers and hides in the Arctic and on Gotland (Figure 3.7) we tracked the coming and going of individually marked geese and the establishment and fate of their nests. From these observations we could identify several bird classes, including (i) non-breeders, which did not make a nest, (ii) failed breeders, which produced a clutch but lost all eggs during the process of incubation and (iii) successful pairs, which hatched at least one gosling. Successful pairs were also observed throughout the brood-rearing period, enabling us to determine any change in brood size. Clutch sizes were usually recorded in the middle of the incubation period. Successfully incubating birds required approximately 30 days for egg-laying and incubation (Prop & de Vries 1993). The last chance to determine brood sizes in Svalbard was immediately before families migrated south, when they aggregated at well-fertilised snow-free slopes under seabird cliffs.

In the main colony on Gotland, hatch dates were estimated from observations of marked families leaving their nests or by estimating the age of newly hatched chicks (less than two weeks old) and then backdating (Larsson & Forslund 1991). The number of fledged young was defined as the number of goslings observed on the day closest to 20 July (within ± 14 days), which was one to two weeks before fledging. A pair was defined as successful if it produced at least one fledged young. Post-fledging survival – that is, survival from fledging up to the arrival on the wintering grounds – was generally high (approximately 92%; van der Jeugd & Larsson 1998). The measurement of the number of fledged young used on Gotland can, after adjustment for the post-fledging survival rate, be said to correspond to the reproductive success measure obtained on the wintering grounds for the Svalbard geese (see below).

Partial clutch predation during egg-laying or early incubation was not detectable with the methodology used, so clutch sizes may have been underestimated. Another source of bias comes from individuals that are hosts or victims of intraspecific nest parasitism (Choudhury *et al.* 1993, Forslund & Larsson 1995, Larsson *et al.* 1995, Anderholm *et al.* 2009a,b). In addition, observations of young broods in the breeding colony and on brood-rearing areas show that brood mixing events also occur when goslings go astray and are adopted by adjacent broods (Choudhury *et al.* 1993, Larsson *et al.* 1995). In Chapter 7 we describe the occurrence of these extra-pair phenomena and discuss some of their consequences.

Reproductive success: winter brood size

Most parents and offspring migrate as a group back to the wintering grounds and continue to associate for at least four to six months (Chapter 6). By counting the number of goslings associated with individually marked parents in winter we obtained an annual measure of reproductive success for Svalbard birds. These assessments were based on multiple leg-ring readers' judgements of whether goslings were in close proximity to a marked bird. Our cut-off for making these assessments was 1 January each year, based on the timing of family break-up. Within the flock, family members walk a similar pathway, and coordinate vigilance bouts, aggressive encounters and social displays. Juvenile plumage is distinguishable because of differences in colour on the cheeks, neck, back, wings and flanks. Juvenile feathering is often less well defined than in adults; the whites are not as white and the blacks not as black. In addition to differences in plumage, juvenile behaviours are conspicuous. Even in winter, the nearly full-grown juveniles still produce gosling-like vocalisations and frequently display submissive 'greeting' postures when adults approach.

An unknown number of pairs will have produced surviving goslings that we could not detect with this method. The fact that there were a variable number of 'unattached' single goslings on arrival on the wintering area each year (Chapter 6) indicates the nature of this bias. However, we take advantage of the fact that the vast majority of pairs are still associating with at least some of their goslings during and after autumn migration, allowing us to assign *minimum* reproductive success values for hundreds of pairs each year.

Moult

To establish when Arctic geese moulted in relation to the date of hatch we noted when marked individuals with known breeding history lost their flight feathers. We also kept track of the dates when birds regained their ability to fly later in summer. Non-breeders, together with early failed breeders, moulted on lakes, where they aggregated a few days before the initiation of the moult. Progress of moult was determined by daily counts of the number of flightless and flying birds. Assessing which individuals were able to fly was facilitated by frequent disturbances caused by Arctic Foxes. For each year, median dates of the beginning and end of moult were calculated (Prop *et al.* 1984, 2004).

Measuring goose behaviour

We quantified goose behaviour with a variety of sampling methods depending on specific objectives. For example, we conducted (i) flock scans with instantaneous recording to determine the proportion of the day that was devoted to foraging and other activities, (ii) behaviour sampling during continuous flock scans to quantify the rate of aggressive interactions and other conspicuous behaviours, and (iii) focal animal samples with instantaneous and continuous recording of a variety of events and activities. All of these techniques are nicely described in *Measuring Behaviour* by Martin & Bateson (2007). During a typical focal animal sample a single goose, or in some cases a family of geese, was watched until it went out of view, usually resulting in a 5–10 minute observation session when we quickly and succinctly described all activities as they occurred (head-up, down-graze, preen, etc.) into a hand-held recorder. We later transcribed tapes into digital format using event recorder hardware and software that resulted in timed sequences of each activity, including rate (number per minute), bout length (in seconds), proportion of activity in the total observation time, and inter-bout interval (time between bouts).

We also kept track of the outcome of aggressive encounters for individually marked birds and a value for proportion of wins was calculated. A win was scored when an opponent gave way, thus losing its immediate space and resources in that space. In some of our studies aggressive encounters were ranked according to intensity, ranging from encounters that were won without much more than a look in the direction of a victim to full-scale bouts of biting and wing beating. Other social behaviours among pair and family members were recorded during observations of wild geese and in carefully designed experimental set-ups with captive geese at WWT Slimbridge (Chapters 4, 6).

Molecular genetic analyses

To answer questions about genetic relatedness between socially interacting individuals and genetic structure of populations, various molecular genetic techniques were used. In cooperation with colleagues at genetic laboratories we employed DNA fingerprinting, microsatellite analyses and DNA sequencing (Choudhury *et al.* 1993, Larsson *et al.* 1995, Anderholm *et al.* 2009a,b, Pauliny *et al.* 2012, Jonker *et al.* 2013). DNA was isolated from tiny blood samples, usually less than 100 microlitres, which we collected during moult catches. We sampled the blood from wing veins of the geese. Blood was stored in alcohol or buffer in freezers until analysis. Blood sampling of wild birds is nowadays a rapid and well-established standard procedure and does not negatively affect the birds' future life.

To analyse the frequency of intraspecific nest parasitism we also used non-destructive egg albumen sampling and protein fingerprinting. Albumen proteins are genetically variable among females and their electrophoretic band patterns are useful both for identification of parasitic eggs and for estimation of host–parasite relatedness. An egg can be sampled up to a week after being laid. The egg albumen was sampled through a drilled hole at the narrow end of the egg. The hole was thereafter sealed with cyanoacrylate glue. The sampling process does not affect egg hatchability (Anderholm *et al.* 2009a,b).

Assessment of diet

Geese are almost completely vegetarians, although they probably do not pass up the odd worm or insect when encountered. For this reason goose biologists spend a lot of their time studying plants – goose food – as well as the geese. The first task is to determine which plants foraging geese take.

We used two methods for determining goose diet, one direct and the other indirect. When we could identify plant species from a distance it was possible to record the number of pecks directed toward each food item. This direct method is particularly useful for geese on incubation breaks that forage on obvious patches of distinct food types. The observer has to be near the focal animal to make use of this technique.

When it was not possible to observe the geese directly or when they foraged in a mixed sward, diet was assessed by identifying items based on cell wall structure of plant fragments found in droppings, or faecal stools (Owen 1975). We collected droppings from different habitats and in different seasons. Diet analysis by examining droppings is possible because around 70% of ingested food was passed through the gut in a largely undigested form and because each plant species has a unique cell wall structure when viewed through a microscope (Figure 3.8). Samples were stored after drying overnight. Box 3.1 describes this method of diet analysis in more detail.

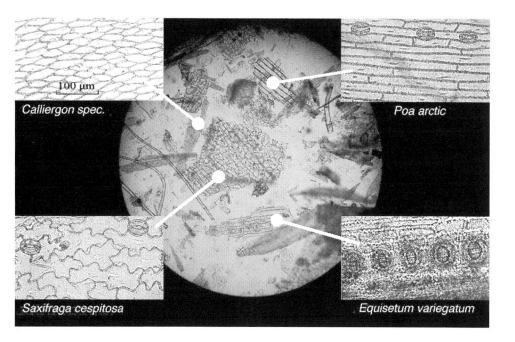

Figure 3.8. *Diet composition as assessed by microscopic analysis of droppings. Viewing droppings through a microscope reveals the remains of plants ingested by the geese (centre of the picture, magnification 100×). The cell wall structures encountered are compared with a reference collection (inserts show some example species). The example dropping was collected in Svalbard, and contained a mixture of mosses (Calliergon), grasses (Poa), herbs (Saxifraga) and horsetail (Equisetum). Photo by Christiane Hübner.*

BOX 3.1 DIET ANALYSIS BY EXAMINING DROPPINGS

This box contains information about determining diet based on the undigested plant fragments in goose droppings (faeces).

Once geese left an area, fresh droppings were collected in bags, taking care to record quantity. We usually collected samples of 25 droppings, although in some situations it was possible to retrieve individual droppings from marked birds with a known age and status. Care was taken to collect only droppings originating from the focal habitat, achieved by waiting for at least 1.5 hours, the usual throughput time for goose food. In other words, only droppings produced 1.5 hours after the birds' arrival will contain material from the focal habitat, while earlier droppings will have originated from a previous foraging location. Therefore, we kept track of time and the location of the geese as they travelled through the habitat. Droppings were stored after drying overnight in an oven at 70°C.

Droppings were examined under a microscope following methods described by Owen (1973), but modified as follows: (i) to homogenise the sample, droppings were ground through a sieve (2 or 5mm mesh width), (ii) a small amount was thinly spread and mounted in water, and (iii) the slide was systematically sampled every 5mm, along transects 5mm apart, at a magnification of 100–400× until 100 fragments were assessed.

Identification of the fragments was based on the specific structure of the epidermal cells, which is unique for each plant species. Identification was made possible by preparing a set of 'known' epidermal sections from plants collected from the birds' habitat. Thin slices from upper and lower sides of the leaves were prepared and mounted on microscope slides. Single-layered leaves of mosses and the thin petals of flowers were mounted directly. It was helpful to prepare a photographic library of specific cell walls to display in the lab. On occasion, in the interest of time, we identified fragments to the level of genus, or even broader categories, e.g. mosses, grasses, herbs. After samples from different vegetation types were grouped in periods (e.g. five days) we calculated a 'crude' average diet composition through the season.

In order to calculate a more accurate measure for birds that spend variable amounts of time in different habitats eating different types of food, one can include a factor for relative densities of droppings in the field and another referring to the size and weight of different food items, as explained by Prop & Deerenberg (1991).

Grazing pressure (dropping densities)

In order to compare grazing pressure between vegetation types we recorded the number of droppings per area that accumulated over time. We typically placed at least 10 circular plots (each with an area of 4m²) in each habitat type. Every one or two days, droppings in each plot were counted and removed. Longer intervals between counts were possible at times without much precipitation. Weights of droppings varied widely between vegetation types, but the time to produce a dropping (dropping interval) was relatively constant for a given seasonal period (Prop & Black 1998), which made dropping densities a reliable measure of grazing time spent in different areas and habitats.

BOX 3.2 PHENOLOGY AND DEPLETION OF FOOD PLANTS

The term phenology here refers to the emergence, subsequent growth and senescence (death) of plants. We were interested in the fate of each food item that was available to the geese and in how much of the plants were removed by the birds. In habitats that were heavily grazed and regrazed by the geese it was necessary to temporarily inhibit the birds' access to the plants to be able to measure daily plant growth, because otherwise geese would remove the new growth before we had a chance to measure it. We were also interested in quantifying the growth rate of plants that were not being grazed at all so we could compare these with grazed plants that were investing in compensatory growth after each grazing event.

A typical plot design, therefore, included groups of four 20×20cm quadrats that were established 2m apart and marked with small numbered sticks. One quadrat was initially enclosed by a thin mesh fence (e.g. chicken wire or netting), while a second was temporarily left open allowing access by the geese, a third was permanently enclosed for the duration of the study period, and a fourth was permanently available to goose grazing. We swapped the wire fence between quadrats one and two on four-day intervals but the fence around the third was left in place. The geese did not seem to be shy of the short fences (c. 50cm in height) because they grazed the grass right next to them, but they were inhibited from reaching inside the enclosures.

Within each plot, 12 different shoots of grass were marked with tiny plastic coloured rings, in order to follow the fate of individual plants. We accomplished this by cutting plastic drinking straws into small rings (c. 2mm tall) that were snipped in the middle so they could be opened and closed around the base of a plant. Coordinates of the marked plants within the plot were noted to facilitate finding the small rings. Marks were replaced if they were lost. At intervals of one to four days every shoot was examined to record the following parameters (Prop 1991):

- Length of each leaf to the nearest mm: for the youngest leaf (leaf 0) this is the distance between the leaf tip and the ligule of the upper fully emerged leaf (leaf 1); all other leaves were measured as the distance from the tip to the ligule (see Figure 3.9).
- Status of every leaf: a) grazed or not; b) the colour of the leaf blade: fresh green, with a brownish tip (partly dead), or completely brown (dead).
- Number of daughter tillers.

Based on records from one sampling occasion to the next the following parameters were calculated to define growth of plants and the intensity of grazing:

- Growth rate per shoot as the sum of the extensions of the leaves, which in grasses is restricted to the two youngest leaves (leaf 0 and leaf 1).
- Birth rates of leaves as the interval in days between appearances of successive leaves on the same tiller.
- Death rate of leaves as the number of leaves that die per tiller.
- Amount of leaves removed by grazing.
- Age of a leaf at the moment of grazing, which is an indicator of the quality of food.

All measurements of length of the food plants were converted to weights by using calibration curves based on the relationship between length and dry weight of leaf blades.

Figure 3.9. *(A) Measuring individually marked shoots to determine growth and depletion of goose food plants (Jeff Black). (B) Photo of individual shoot of* Puccinellia maritima *with ring at the base. The youngest leaf (L₀) is at the top of the shoot; the older, full-grown leaves (L₁ and L₂) occupy lower positions. The oldest leaf (L₃) of the example shoot has died (Jouke Prop).*

Production and consumption of food plants

Understanding diet composition and use of habitats depends on the quantification of food availability. Sampling over spatial scales was achieved with a variety of methods, including the use of a 10-point sampling frame within a mixed sward, percent cover assessments inside plots positioned in different habitats, and point-intercept transects through habitats (Higgins *et al.* 1996). Each of these techniques allowed us to quantify the variety and relative abundance of goose food types. By reviewing other habitat features we could describe potential reasons why goose food varied among microclimates and habitat types, e.g. nearness to the shore, elevation gradients, or farming regimes (e.g. Owen *et al.* 1987). The change in availability of goose food over time was determined by repeatedly measuring individual plants (Figure 3.9), as explained in Box 3.2.

In some studies the amount of food removed by geese was determined back in the lab by measuring 3D photographs that were taken of the plots at one- to four-day intervals, and on some occasions twice in one day (before and after a visit by a goose flock). A stereo camera was used and successive photographs compared using a stereoscope or appropriate software (Prop & Loonen 1989, Prop & Deerenberg 1991).

Food intake

To compare foraging performance in different habitats we quantified various aspects of the birds' behaviour with the aim of calculating individual intake rates, which is the amount of food ingested per time. Intake rate was determined directly from observations of foraging behaviour or indirectly by keeping track of dropping production. Box 3.3 describes both methods in detail.

BOX 3.3 INTAKE RATE

Method 1: Instantaneous intake rate

Intake rate (IR, reported in grams per minute) was calculated as the product of peck rate (P, number per minute) and bite size (B, reported in grams after drying): $IR = P \times B$.

Peck rates were estimated by recording the time required for 50 pecks and converted to number per minute. Pecks were only timed as long as the goose was feeding, and records were aborted as soon as feeding was interrupted by a short spell of vigilance. When possible, we noted the food plant type that was being eaten. This was difficult when geese were foraging on a short, heavily grazed sward, but became more relevant and feasible when they were on heterogeneous vegetation zones or when separate plant species were easily recognisable. Such vegetation 'patches' were common in natural habitats such as saltmarshes and in Arctic regions. Observations were facilitated by a good hide, preferably fixed on a tower a couple of metres above the ground, or positioned on a rocky outcrop.

We determined the amount of food that was taken in a bite and then collected, dried and weighed similar sized items. This was straightforward for discrete buds, flowers and seed heads, or when geese selected whole plants, like seedlings in spring or small Arctic herbs in summer. However, when the geese fed on grasses, bite sizes were more difficult to determine. This was because one has to consider that they usually selected the tips only, or took more than one leaf in a single bite, as explained below.

An accurate way to determine size of leaves taken was to compare the length of marked plants before and after a grazing event (see above). The advantage of this method is that information on growth and death rate of grass shoots is obtained. The disadvantage is the large investment in time required for sampling. We also employed two alternatives: (1) through time, plots were

Figure 3.10. *Geese move fast and peck at high rates. For detailed assessments of foraging behaviour hides were required to observe the birds at close quarters. The picture shows a goose grazing in a marked plot of 20×20cm. Photo by Jouke Prop.*

photographed frequently (3D photographs of 20×20cm plots have proven to give sufficient detail), (2) in other situations, we estimated the amount of grazed material by comparing with adjacent leaves (grazed and ungrazed) with similar widths. The difference between the two provided an estimate of leaf lengths that were removed by grazing. Each of the methods requires calibration curves in order to convert lengths into dry weights.

The number of leaves removed per peck was assessed by close inspection of the sward, which allowed detection of separate 'bite-sites' containing one or more adjacent leaves with fresh bite marks. The reliability of this method was determined by examining plots after they received a particular number of pecks. This was achieved by watching an easily observed plot and counting all pecks that were taken in the plot as geese passed through (Figure 3.10). The total number of leaves removed divided by the total number of pecks provided the estimate of leaves taken per peck. Gross intake can be converted to net intake with a further step (see below).

Method 2: Intake rate by dropping production

The rate of ingestion of food was calculated as:

$IngestR = (W/I) \times (100/(100 - D))$

where W is the average dropping weight, I is the average dropping interval and D is the digestibility of the food. By considering digestibility the rate of droppings produced is converted into the ingestion rate. Because geese ingest sand to facilitate the mechanical destruction of the food, dropping weights are best expressed in terms of ash-free dry weights. The quotient of W and I estimates the egestion rate during foraging, and observations of dropping intervals were therefore only collected while geese were active (feeding plus short vigilant spells, as opposed to periods of resting/preening lasting for at least several minutes). To obtain an estimate for the intake rate of food, ingestion rate was corrected for the intensity of feeding, or the proportion of time that was actually spent feeding (*FT, feeding time*) during foraging (*ForT, foraging time*):

$IR = IngestR/(FT/ForT)$

FT was recorded during focal observations using a stopwatch that was turned on when the bird was foraging and off when it was not foraging. Foraging included moments when a bird was looking and approaching food items with an angled neck and lowered head and when food was manipulated in its bill. *ForT* was the total time of the sampling session (feeding plus short vigilant bouts). For method 1 and method 2 we estimated digestibility D in a variety of ways. First, by using a natural marker in the food:

$D = 100 \times (1 - M_f/M_d)$

where M_f and M_d are the concentrations (ash-free) of the marker in the food and droppings, respectively. The toughest parts of cell walls, which are not easily digested, are appropriate marker substances (estimated following the acid detergent fibre, *ADF*, procedure). For a proper comparison, M_f was weighted by the relative importance of each food species in the diet (see diet analysis Box 3.1).

In cases when fibre content of droppings was not available, digestibility was estimated from the concentrations of either nitrogen (N) or ADF in the food plants. Based on a large number of samples collected on several goose species the required regressions are:

$D = 15.04 + 6.25 \times N$, or $D = 56.62 - 0.75 \times ADF$ (Prop & Vulink 1992)

The digestibility of food depends on the chemical composition of the food, and on the retention time in the digestive tract (Prop & Vulink 1992). The formulae above are average estimates for the winter and spring. They may give biased results in summer because at that time of the year retention times are much longer. Most reliable estimates can be obtained by allowing for food composition and food retention time. Dropping intervals (I, minutes) are positively related with retention times, based on digestion trials with captive Barnacle Geese (Prop et al. 2005). Therefore, digestibility of food, adjusting for retention time, can be estimated by:

$D = 56.7 + 3.11 \times I - 1.64 \times ADF$

Net intake rates NIR were derived from gross intake rates, digestibilities and proportional ash content Ash as:

$NIR = IR \times Dig \times (1 - Ash)/100$

Statistical tests

We employed an ever-changing series of statistical approaches. At the beginning of the study, analyses were often done by hand with calculators or with long-winded computer code to accomplish simple analytical tasks. Over the years we benefited from the development of commercial statistical computer programs, and data presented in this book have been analysed using a wide array of packages, including successive versions of GLIM, MARK, SAS, SPSS and R. As the power of the personal computer increased so did the rigour in statistical testing. We agree with Brown & Brown (1996), authors of a commendable long-term study on Cliff Swallows *Hirundo pyrrhonota*, who state that a conservative approach emphasising relatively simple and straightforward univariate analyses is desirable. In keeping with this preference we frequently used nonparametric tests, thus avoiding potential problems of unmet assumptions that can accompany parametric tests. Where circumstances allow, however, we also employed multivariate models to better understand the strength of relationships among variables. Analyses of variance and covariance were performed with appropriate procedures, including general linear models (GLM). Sums of squares of an effect were usually adjusted for any other effect in the model (type III sums of squares). Multiway contingency tables were analysed by general log-linear models (GLLM). We followed the traditional model selection, which is based on comparing measures of variance explained by models with and without a variable of interest. Variables were dropped from a model when the lower variance explained by the reduced model was offset by a higher degree of simplicity, thus leading to the most parsimonious model. The decision to include a parameter or not was determined by the outcome of the appropriate test, for example an analysis of variance or a likelihood ratio test. Depending on the distribution of the test statistic, we report in this book the associated F or χ^2-value with its degrees of freedom and assessed against $\alpha = 0.05$. In line with a general trend, and stimulated by the use of the program MARK, we cautiously shifted to a newly developed way of model selection, which is based on information theory with the AIC (Aikaike's Information Criterion) as a tool to assess the support for a model (White & Burnham 1999).

Summary

We employed a variety of field and lab techniques to study the geese while attempting to minimise the possibility of disturbing their daily routines. Annual population assessments of the Svalbard population were conducted on the wintering grounds when the birds were concentrated in relatively few areas. Annual estimates of the size of the Baltic population were derived from counts of nesting pairs. At other times of year we were content to record daily numbers and distribution from vantage points at key stopover sites and breeding areas. We captured most geese during their flightless period by herding them into corrals erected on shorelines. Each bird was fitted with a metal ring and a durable plastic ring engraved with unique alphanumeric characters. The geese were sexed through cloacal examination, and aged according to their plumage colour and patterns. We also weighed and measured each goose, including measures of skull, tarsus and moulting feather lengths. The key to success in this project was our ability to position ourselves in hides and observation towers where we could resight individually marked birds with spotting scopes. The average bird was seen nine times per year. From these multiple observations we determined the fate of the birds' nests, and of broods after hatching and fledging, resulting in an assessment of their annual reproductive success.

Various molecular genetic techniques were used to answer questions about genetic relatedness between socially interacting individuals and genetic structure of populations. Since geese are grazing specialists we also studied their foraging performance. We quantified activity patterns of flocks and individuals by sampling behaviours making use of stopwatches and tape recorders. We determined the birds' diet by identifying undigested plant fragments in droppings (faeces). By counting dropping densities in the field, we assessed grazing pressure on different food types. We quantified plant emergence, growth, death and depletion (amount removed by the geese) with careful measurements of individually marked shoots and inflorescences. The quantity of food that was eaten per unit time (intake rate) was determined by measuring the number and size of goose bites or by keeping track of the number and size of droppings that were produced during foraging sessions. We designed a number of experiments in the field and when appropriate made use of more controlled situations by studying captive geese.

CHAPTER 4

Finding mates

Geese are well known for their monogamous lifestyle, where females and males live in pairs throughout the annual cycle. Since Barnacle Goose partners depend on each other to rear young successfully and mates stay together for life, we might expect selection to favour the careful choice of initial partners. In species with long-term pair bonds the reproductive experience is a team effort, so taking account of only one member of the pair when explaining reproductive performance is presenting only half of the picture. For example, if large females produce more offspring, high reproductive rates may be attributed to large female size. However, if large females pair with males with the best territories perhaps the females' success should be attributed to the males' qualities which enable acquisition of good territories. In this chapter we describe the process and consequences of mate selection in detail.

Background

In winter, goose flocks are composed of an assortment of singles, pairs and family groups (mature goslings associating with parents). On return to the breeding grounds, when goose pairs are busy establishing territories within the colonies, the previous summer's offspring, now almost 12 months of age (referred to as yearlings), are free from family ties. Yearlings do not attempt to breed in their second summer, spending their time exploring and mixing with other non-breeders (van der Jeugd & Blaakmeer 2001). Young non-breeders sometimes venture into the colony but are readily displaced by older pairs that defend 5–10 metres around nests. Yearlings and other non-breeders are usually relegated to the outer edges of the colony and eventually move to adjacent ponds and lakes to begin their moult while the breeders are still incubating eggs (Prop *et al.* 1980, 1984). The yearling summer is all-important because it is the time when some lifelong patterns begin to take shape, beginning with the process of mate sampling in the non-breeder flocks.

Both sexes are involved in the mate selection process, which consists of a series of stages. The male initiates courtship by directing loud calls and neck stretches at a female. The female may either ignore the male's advances or encourage him with particular types of calls, culminating with a Triumph Ceremony and vocal duet (Figure 4.1). The female chooses whether to respond to the male's advances, and the male chooses to direct his attention her way or to try elsewhere. As pair members spend more time together they maintain greater proximity and respond to each other in courtship displays, but either sex may curtail their involvement at any stage along the way. Once the pair bond is established the Triumph Ceremony is still practiced by pair members throughout the year, with a peak in frequency occurring in the spring (Black & Owen 1988).

Figure 4.1. *Social displays in goose pairs and prospective pairs (after Fischer 1965, Radesäter 1974, Black & Owen 1988, Hausberger & Black 1990). The male, who directs a loud call at the female, initiates the display. If the female responds positively, the courtship proceeds to the next stage of herding where the male actively positions himself right next to the female (1). The male may then launch into a mock attack (2) where he holds his neck down low and advances toward a real or imaginary neighbour. He returns to the female sometimes holding his wings out (3). If the female responds favourably they begin calling loudly at each other, making a duet (4), and the male may attempt to grab the female's neck as she lowers her head (5). These last two components describe what is known as the Triumph Ceremony.*

Time and location of pair formation

By identifying the time of year partners were first seen together we could determine where the pair formation event must have taken place. We were able to do this by recording mate status of individually marked birds throughout the annual cycle.

Determining the origin of mates is important for a number of reasons. Halliday (1983) suggested that the reproductive success of a mating pair might be affected not only by their quality as individuals, but also by their degree of genetic and behavioural complementarity. Partner compatibility may be particularly pertinent where ecological conditions vary over the species' natural range, so that positive assortative pairings between individuals adapted to the same local habitat may be favoured (Shields 1984, Greenwood 1987).

The challenge was to resight birds frequently enough to identify the first time they were seen with their long-term mates. The 'pair bond' concept in our study refers to two birds that are consistently seen together even during the non-breeding season. Although more than 90% of birds initiated their first pair bonds before age five, first pairing ranged between one and eight years in males (average 2.5 years, SE 0.05, n = 530) and one and 15 years for females (average 2.6 years, SE 0.05, n = 611).[3]

In cases when partners were already together when we first saw them in autumn in Scotland, we assumed the partnership began in the summer months in the Arctic and that they completed the southward migration as a pair. On review of these dates, we estimated that over half of the population initiated pair bonds during the Arctic summer: 66% of females, and 64% of males (Figure 4.2).[4] This is at odds with information from the exemplary Lesser Snow Goose *Chen caerulescens caerulescens* study in La Perouse Bay, Canada, where it is assumed that most pairs are established during winter (Cooke *et al.* 1995). However, that study did not

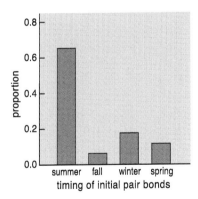

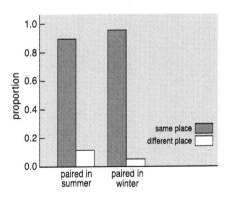

Figure 4.2. *Timing of initial pair bond establishment; data for females is shown. Significantly more birds established partnerships in summer while on the breeding grounds than at other times of year compared to what would be expected if timing of pairing was equally distributed (females, 172 in summer and 90 in non-breeding season). The analysis included pairs where both partners had individually marked leg-rings.*

Figure 4.3. *Origin of mates in relation to timing of pair formation for pairs whose members were captured in their first or second summers in Svalbard (n = 194 paired in summer, 47 paired in winter). In this population, capture location in summer was synonymous with natal origin. The vast majority of mates were from the same capture location; overall average 89%.*

determine pair status in all months and locations so they were unable to entirely resolve the issue. We suspect that the pair formation process begins in the yearling summer for most goose populations, which will have important implications for identifying the origin of mates and movements among colonies (Cooke *et al.* 1995, Ganter *et al.* 2005, see also Rodway 2007).

In an attempt to identify the origin of mates we checked whether partners had been captured previously at the same location as goslings or yearlings, which would indicate that they spent their first and/or second summers together in the same set of ponds and habitats on the tundra landscape. We had this kind of information for 241 pairs where both pair members had plastic leg-rings. We found that for 215 pairs (89%), both partners were previously captured on the same day and location (Figure 4.3). This means that many partnerships comprise birds that grew up making use of the same Arctic 'neighbourhoods' during their first or second summers.

Surprisingly, a proportion of pair bonds that were established later in the large winter flocks 3,100km to the south also consisted of partners from the same Arctic sites. In an earlier analysis, Owen *et al.* (1988) found that 58% of the females from the 1976 cohort eventually chose mates from the same natal origin even though they only had a one in five chance of doing so, assuming that most birds from different colonies mix together in the wintering flocks.

The above observation regarding pairs that are established on the wintering grounds raises an interesting problem. How do so many of the partners from particular Arctic neighbourhoods manage to find each other among the masses of conspecifics that are mixed in large wintering flocks? And why would Barnacle Geese evolve mate choice mechanisms that limit genetic and social mixing?

Mate choice theory

In geese, as in many animal systems, it is assumed that each sex may evolve a particular set of traits that corresponds to the other sex's preferences for those traits, and individuals within each sex compete to display the best that they have to offer. Through this *sexual selection* process individuals with the most competitively successful and the most attractive traits will be favoured by the other sex and gain more matings (Andersson 1994).

Selecting a mate is a complex process involving searching, sampling, information gathering and decision making. With no restrictions on time, movement or memory, an animal's best strategy of mate choice would obviously be to inspect all available partners and then to select the best one. However, animals are usually subject to one or more constraints and are unable to sample all available individuals. Several models have been proposed as to how animals may go about choosing mates under given constraints (Choudhury & Black 1993). *Random mating* suggests that there is no mate choice and that animals settle with the first available mate encountered (Janetos 1980). The *fixed-threshold strategy* predicts that animals will sample mates until they encounter an individual that meets some minimum requirements or threshold value (Janetos 1980, Wittenberger 1983). In the *one-step-decision process*, an animal has to decide at each encounter whether to accept or reject the potential mate. It cannot go back to a previously rejected individual once a decision has been made. The *sequential comparison tactic* also predicts that mates are sampled in sequence, but the choosing animal always compares the two most recent candidates according to some rule (Wittenberger 1983). In the *best-of-N-mates strategy* an animal examines N potential partners, ranks them on a relative basis, and chooses the best one; N may be the maximum number an animal can sample within the given time and space or within its memory capacity (Janetos 1980, Wittenberger 1983).

In order to identify how Barnacle Geese deal with the problems of mate selection, for two years Sharmila Choudhury closely monitored the early life associations of 78 young geese that were raised under controlled conditions at the WWT Slimbridge avicultural facility. Her results are developed in the next sections, followed by further information about the outcome of partnerships in wild geese.

Mate choice experiment

Incubator-hatched goslings were placed in eight unisex rearing pens equipped with ample water, food and shelter. When the geese were 17 months of age, groups of males and females were placed in two large enclosures and allowed to pair up freely until the end of the following summer, when most new pairs establish nests and lay their first clutch of eggs. We measured several potential mate choice characteristics that could be used to distinguish the birds' phenotype (e.g. body size and cheek-patch plumage patterns). We also quantified the birds' relative dominance rank, vocalisation rate and vigilance behaviour after presentation of a stuffed fox 'walked' past the enclosure. These traits may be used as potential cues about a bird's ability and willingness to invest in protection of mates, nests and offspring. Pair formation behaviours and the identity of birds involved were initially recorded on a biweekly basis and then each morning in the three months prior to nest initiation.

Table 4.1. *Sequences of mates sampled by Barnacle Geese during mate choice (for birds sampling more than one partner). A hyphen connecting birds indicates a partner-hold strategy, i.e. sampling more than one potential partner at a time (a-b = partner-hold strategy result, animal moved from a to b; a-b-a = partner-hold strategy result, animal stayed with a). Asterisk (*) indicates final mate (some males were polygynous). From Choudhury & Black (1993).*

Males	Mate sequence	Females	Mate sequence
BVE	LSS LZS-LSV*	LZS	LPA BVE BEF LPC*
BEF	LZS-LXI-LZS LZS-LSV-LZS LSZ-LSS*	LSS	BVE PAB BEF*
DEA	EEE-DEK DEK-LLU LLU-LYT-LLU*	LZX	LYU-KAP-LYU*
DEG	LJA-LXC-LJA*	EEE	DEA NAB*
LPA	LZS-LSP*	DEK	DEA-LSJ-DEA NAD*
NAB	LYT EEE*	LXI	BEF PAB*
PAB	LSS-LXC-LSS LSS-EEV* EEV-LXI*	LXC	PAB DEG
PAF	LPY* LPY-PAH*	LSV	LPT-BVE BVE-BEF-BVE*
LTF	BJT-LLU-BJT*	LLU	LTF DEA*
KAP	LIS-LZX-LIS*	LYT	DEA NAB
LYU	KHR* KHR-LZX*	LXN	KKC ECE*
KKC	LXN EXE LIN* LIN-DEI* DEI-NAU* NAU-LUT*	NAT	LTZ-NAZ NAZ-EIE*
LTZ	NAT-DEF* DEF-BEA*	LIN	LVD KKC*
LVD	LIN LYF*	NAU	LNV KKC*
NAZ	NAT-BEA DJE LVH*	LUT	BUJ-KKC*
		BEA	NAZ-LTZ*

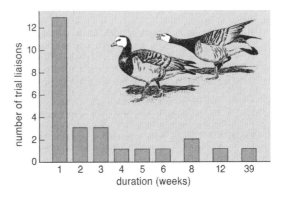

Figure 4.4. *Duration of individual trial liaisons during mate selection in Barnacle Geese. The start of a trial liaison was defined as two birds carrying out the same behaviour in the same direction at not more than one goose length (45 cm) apart and at least three goose lengths from another bird. Behaviours included social displays, vocalisations, leading/following, food sharing, friendly approaches, head-dipping and copulation (from Choudhury & Black 1993).*

About half the birds (51%) settled with the first mate they sampled, whilst the other half went through one to six potential partners or trial liaisons before settling with a consistent mate (Table 4.1). Trial liaisons appeared to be indistinguishable from permanent partnerships except for their temporary nature. Liaison sessions ended when trial partners directed aggressive threats at each other, ignored courtship advances, or started to court a new trial partner. There was no difference in the number of liaisons terminated by males and females (10 each), supporting the view that both sexes are responsible for the mate choice process.

The duration of individual trial liaisons varied greatly from one day to over nine months, but the majority lasted only a few days (Figure 4.4). Trial liaisons early in the season tended to be longer, and became shorter the closer it got to the time of breeding.[5] Birds that started searching for a mate earlier in the season sampled more potential partners than those that started late.[6]

These observations may underestimate actual mate sampling in the wild, because of the limited number of birds in our experiment and because of our classification of trial liaisons. The mate choice process may be far subtler, and some mate sampling and decisions may occur before trial liaisons are formed. Mate sampling could, for example, take place at two levels (Choudhury & Black 1993). Animals might eliminate totally unsuitable or unattractive mates in a first selection round, without forming a trial partnership. The next selection step may involve going through trial liaisons with the potentially suitable candidates. Indeed, pair formation in geese consists of a series of courtship phases, and we observed a number of cases where a male initiated courtship towards a female, but then moved on to a different female before obtaining a positive response. These cases were not included in the study and it is difficult to tell from such brief encounters whether the male terminated the pairing process, or whether the female's lack of response or rejection caused the male to leave. These problems will occur in studies of any species where animals may assess the quality of mates without any recordable sampling behaviour. Thus, we were able only to analyse mate sampling behaviour that occurred above a certain level of interest by the prospective partners.

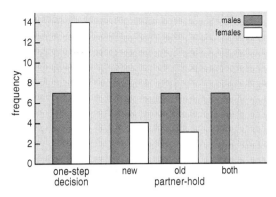

Figure 4.5. *Frequency of sampling strategies used by geese that sampled more than one potential partner. On the x-axis the one-step decision strategy refers to individuals that moved from trial partner to trial partner in a forward, step-wise process. During the partner-hold strategy, birds either moved on to the new trial partner, stayed with the old one, or formed a polygynous association with both trial partners (from Choudhury & Black 1993). Also see Table 4.1.*

Sampling strategies

Figure 4.5 shows the different sampling procedures used by the geese that sampled more than one potential mate during the mate choice process (also see Table 4.1). In about 40% of the cases birds moved step-wise from trial partner to trial partner in the way predicted by the one-step decision process, and since they never returned to a previously sampled mate nor appeared to be using a threshold criterion, it seems unlikely that they were using the best-of-N-mates or the fixed-threshold strategies. In the remaining 60% of cases, the geese appeared to use a modified version of the 'sequential comparisons rule' described by Wittenberger (1983). Instead of leaving one trial partner for the next, they held on to the old partner for a time, whilst sampling the new one. Often this resulted in temporary trios, where three birds moved through the flock in a close unit. A male in this situation would usually direct most of his attention to the new female and return regularly to the old one. Females, however, did not take the initiative in courtship, so when a female was involved with two males she only needed to respond positively to both. The female would either alternate between the two males in approaching and maintaining proximity to them, or she would be courted and herded alternately by the two, both competing for her attention.

This *partner-hold strategy* allows the choosing individual to compare the relative qualities of two mates before making a decision about whether to leave the old partner for the new one. This strategy should result in choosing the best of all those sampled, yet has the advantage of freeing birds from the constraints of remembering the qualities of all mates encountered and of relocating the best mate. This makes good sense especially for a flock-living species where it might be difficult to relocate previously sampled partners.

In conclusion, Barnacle Geese appear to use a combination of the one-step-decision and partner-hold strategies, with females predominantly using the former and males the latter strategy. This may partly be related to the fact that males could often retain the interest of two females for quite a while, even to the extent of forming polygynous breeding associations (at least in captivity where fewer mates were available). However, both sexes seem to be exercising choice simultaneously where the response of the sampled mate seemed to affect the outcome of the encounter. The differences in sampling strategies of the sexes may thus result from differences in choosiness, i.e. the strength of their preferences.

Traits and preferences

Preferred traits are expected to be correlated with an individual's investment in future offspring (Trivers 1972). High-quality birds will be in greater demand as mates and can thus afford to be choosier themselves (Burley 1977), partake in multiple assessments of potential mates (trial liaisons), and find a suitable partner at an early age.

We found that heavy and more vigilant females, as well as those with darker face patterns, had more trial partners (Table 4.2). Since greater body mass means greater fat reserves for breeding, and therefore a better chance of rearing offspring successfully (Choudhury *et al.* 1996), heavier females may be higher-quality mates. We also found that larger females were preferred as mates in the wild population where pairing age ranged from one to 15 years; larger females paired at an earlier age than smaller females (Choudhury *et al.* 1996). Similarly, increased vigilance of the female may reduce the vigilance burden of the male when protecting the nest and young, thus making a vigilant female more attractive as a mate. The role of face patterns in mate choice is subtler; darker face patterns were not correlated with other female traits. Perhaps signalling properties are enhanced with a larger black eye-patch that contrasts against the white face (review Figure 1.4). In our Baltic study population, there was no indication that birds with particular facial patterns produced more eggs or more fledged goslings; however, related individuals had similar facial patterns (Larsson unpubl. data).

In other birds plumage characteristics sometimes serve as a 'badge of status' where the number and size of plumage patches is correlated with dominance, which is usually linked with age (Rohwer 1985). We know little about the role of honest signalling potential in waterfowl plumages; perhaps this is a valuable avenue for future research (see Torres Esquivias & Ayala Moreno 1986, Kristiansen *et al.* 1999). In Black-capped Chickadees

Table 4.2. *Relationship between female and male characteristics and number of potential mates sampled. From Choudhury & Black (1993). * Pearson correlation coefficient. NS (not significant); P > 0.05.*

Character	Females			Males		
	r*	n	P	r*	N	P
Weight	0.377	39	< 0.02	−0.033	38	NS
Abdominal profile	0.419	23	< 0.05	0.079	23	NS
Size (PC1)	0.204	39	NS	0.060	38	NS
Vigilance 1	0.697	12	< 0.02	−0.251	12	NS
Vigilance 2	0.570	12	< 0.05	−0.046	12	NS
Dominance	0.076	23	NS	0.018	23	NS
Vocalisation rate	0.016	39	NS	−0.139	39	NS
Face pattern (% black)	0.366	37	< 0.05	−0.109	39	NS

Poecile atricapilla the contrast between reflectance qualities of white and black plumage patches was greater in males than in females and high-ranking males that are preferred by females as social partners exhibit significantly darker black plumage than low-ranking males (Mennill *et al.* 2003).

For males none of the traits investigated was related to the number of trial partners (Table 4.2), but five males that managed to form lasting partnerships with two or more females right up to and through the breeding season were more vigilant (prior to pairing) after the appearance of a predator than other males.[7] Females may benefit directly from male vigilance. Teunissen *et al.* (1985) found that female Brent Geese paired to more vigilant males enjoyed enhanced foraging opportunities.

How do birds find partners from their own Arctic neighbourhoods?

Earlier in this chapter we showed evidence that about 89% of young geese paired with individuals from their own Arctic neighbourhoods. However, about 35% of the birds paired with their mates during the winter when individuals from Arctic colonies intermingled. How do Barnacle Geese recognise Arctic neighbours within the large wintering flocks? Two mechanisms could underlie the observed assortative mating pattern: (i) birds may recognise previous associates and preferentially choose familiar individuals from the breeding grounds when encountered in the wintering flocks; or (ii) birds from different breeding areas may share particular phenotypic traits, and geese may choose mates with phenotypic traits common to their own area even when they never had the opportunity to associate and become familiar with each other on the breeding grounds. It is clear that locating members of the same breeding colonies within wintering flocks of thousands of birds requires some form of recognition.

We designed a two-part experiment to test whether recognition and mating preferences for colony associates was based on familiarity due to prior association, or on a phenotype matching mechanism (Choudhury & Black 1994). In conjunction with the mate choice study above, incubator-hatched goslings were raised in unisex groups of 11–12 birds each (Figure 4.6A). We tested for the effect of familiarity – i.e. early life association – by allowing two groups of males and females limited access to one another during the first year of life. The unisex groups were allowed to mix with their 'visual' neighbours for short intervals, simulating time together during their yearling summer in the Arctic. At an age similar to wild birds' second migration to the wintering grounds we placed all four groups together and allowed them to pair up freely until the second summer when most new pairs established nest sites in the enclosures. In the second experiment we tested whether geese would preferentially choose mates from their own natal population if they had never met before. In other words, would geese choose mates based purely on some phenotypic cues that were common to their own breeding stock, such as special plumage types, vocalisations or other behavioural characteristics? For this second experiment we obtained eggs from two different stocks and reared them in four separate unisex groups (Figure 4.6B). For the first 16 months of life none of the geese were allowed to see or spend time with geese from other groups. As with experiment one, all birds were moved into one large enclosure at the beginning of their second winter, and allowed to pair freely until the following summer.

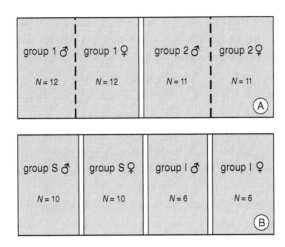

Figure 4.6. *Experimental set-up for mate choice experiments with captive Barnacle Geese. (A) Experiment 1: testing the effect of familiarity – i.e. early life association – on future mate choice. The dashed line indicates a wire fence allowing males and females to see and become familiar with each other. (B) Experiment 2: testing whether geese can distinguish between birds from two different populations by phenotype; Slimbridge (England) and Islay (Scotland) semi-captive stocks, marked as S and I, respectively. The double striped lines indicate solid fences, which did not allow visual contact. For each experiment, all males and females were placed in the same enclosure at the age of 17 months and allowed to pursue trial liaisons and final partnerships prior to and during the nesting season (from Choudhury & Black 1994).*

After watching the birds during the pair formation process we found that birds in experiment one had more trial liaisons with familiar than unfamiliar individuals,[8] and eventually paired with familiar individuals more than would be expected by chance[9] (Figure 4.7A). In experiment two, pairing was random with respect to population and/or genotype; i.e. birds did not prefer mates from the same genetic stock either in mates sampled[10] or in their final partners (Figure 4.7B).[11] The experiment showed that social associations in early life may well be responsible for shaping future mate preferences. The geese did not appear to discriminate between close and distant colony mates purely by phenotypic characteristics.

Choudhury & Black (1994) also found that geese choosing from a set of familiar individuals found partners sooner than those choosing from unfamiliar individuals.[12] Pairing early may have numerous advantages, for example by (i) allowing ample time to establish a workable relationship with a mate prior to the breeding season, (ii) increasing social status and access to food sources, and (iii) ensuring a better choice of mates when the potential mate pool is still large. Pairing with a familiar mate may also improve the chances of success in reproductive attempts since both members of a partnership will have already built a base of experience in a particular breeding area. This idea has yet to be tested but it seems possible since learning the terrain and location of food and predators may be essential before reproductive efforts yield success (Chapter 12).

We also considered whether the birds in the mate choice experiment used other characteristics in addition to the familiarity of potential partners (Choudhury & Black 1994). In this case we used timing of pairing as a measure of mate preferences based on the assumption that high-quality mates may be in greater demand and get taken out of the pool of potential mates sooner than low-quality mates. We found that larger and more

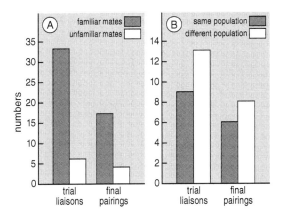

Figure 4.7. *Frequency of trial liaisons and final pairings between (A) familiar and unfamiliar geese, and (B) those from the same and different populations. We reared 78 goslings that were hatched in incubators for this experiment. Eggs were taken from the feral flocks at Slimbridge (England) and Islay (Scotland) (from Choudhury & Black 1994).*

dominant males paired earlier.[13] Males with larger structural size may be able to acquire and store more nutrients and fat, and hence invest relatively more in their mates and offspring (Ankney 1977, Choudhury *et al.* 1996). None of the female characteristics were correlated with timing of pairing.

Pair bond characteristics

Extensive observations at WWT Caerlaverock and other locations allowed us to identify partner characteristics and partnership fates for 946 pairs where both the female and male pair members had been fitted with engraved, colour leg-rings. Figure 4.8 shows ages for partners in initial and subsequent pair bonds. It is remarkable that few young birds paired with older partners; only 5 of 113 first-time pair bonds consisted of different aged mates.

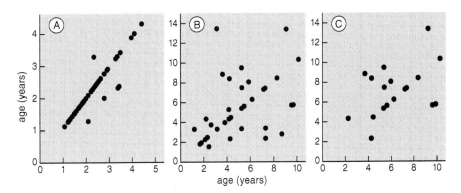

Figure 4.8. *Age of partners at time of pairing (in years). (A) First-time pairings: both partners had no previous mates. (B) Second plus pairings: at least one partner had one or more previous mates. (C) Second plus pairings: both partners had one or more previous mates. From Black & Owen (1995).[14]*

After loss of an initial partner, re-pairing could be rapid; 13 geese established new pair bonds within 20 days. However, it took most geese three to nine months to re-pair with new mates after an initial mate died or disappeared (Owen *et al.* 1988). Although ages did not match as closely as in first-time pairs, the ages of replacement mates were also closely correlated (Figure 4.8A compared with Figure 4.8B, C). Partners were exactly the same age in 46% of these second pair bond cases; in two cases 12-year-old birds re-paired with each other. When partner ages did not match, the older sex was only slightly skewed toward females (57% older female, 43% older male). In four cases, the disparity in age was 15 years.

Examination of 39 replacement mates where both pair members had detailed records provides further evidence of preference for familiar partners (Black & Owen 1995). In 18 cases both members of the new pair bond were older birds that had lost their original partners. In 14 cases the new partners were previously seen on the same coastline in the Arctic, indicating that they could have been acquainted before they became paired. In eight of the cases the new partners had actually been captured together years before, when they

Table 4.3. *Case histories of re-pairing for eight known-age male and female geese, and the number of young, unpaired birds in the population. After the loss of their previous mates, these birds formed new partnerships in their 4th to 10th year with partners that were exactly their own age, partners with whom they had associated during their gosling or yearling year, known because they were captured and ringed in the same brood-rearing areas in Svalbard. From Black & Owen (1995).*

Pair	Mate number M/F	Age at re-pairing (years) M/F	Date of re-pairing	Time taken to re-pair (months) M/F	Cause of previous split[a] M/F	Place of re-pairing	Number of young unpaired geese[b]
LIY LIV	2/2	5/5	Jun 1981	4/6	MD/MD	Scotland	1,830 1 in 3.8
*DY DTP	2/2	6/6	Feb 1982	16/3	UPD/SPL	Scotland	1,721 1 in 4.3
CLV DIP	3/2	7/7	Nov 1983	0[c]/7	MD/UNK	Scotland	995 1 in 7.8
*DY DUK	3/2	7/7	Oct 1983	8/23	MD/MD	Svalbard	1,127 1 in 6.9
$PH XJU	2/2	4/4	Dec 1984	1/0	UNK/ UPD	Scotland	1,056 1 in 7.4
CZP CRP	2/2	8/8	Nov 1984	5/6	UPD/ UPD	Scotland	900 1 in 8.6
YRP YFK	2/2	6/6	Sep 1984	7/6	UPD/ UNK	Svalbard	1,204 1 in 6.5
XCC XAH	2/2	10/10	Oct 1990	5/8	UNK/ UNK	Svalbard	1,276 1 in 8.3

[a] Codes for status of previous mate: SPL Split with previous mate, MD Mate disappeared/died, UPD seen as unpaired prior to re-pairing, hence could have split with previous mate or the mate could have died – mate was unringed, UNK Unknown status, mate was unringed.

[b] Calculated from data in Owen *et al.* (1988) (e.g. the timing of pairing for young birds is 8% by age 15 months, 19.2% by 21 months, 30.1% by 27 months and 42.7% in later months) and unpublished demographic data.

[c] Mates overlapped in summer months, i.e. one male and two females. This male, CLV, had three different polygynous trios in his 11-year lifetime – a rare occurrence.

were goslings or yearlings (Table 4.3). This means they spent their first and perhaps second summers together in the same Arctic neighbourhood. Birds involved in these cases of re-pairing at an older age apparently preferred similar aged partners and avoided younger birds that were available in the population. One female (yellow DUK) whose initial mate died was subsequently unpaired for 23 months even though there were over 1,000 younger potential mates available in the population (see below). She re-paired at age seven with a mate with whom she shared the same brood-rearing areas in their gosling summer. Pairs often return to the same foraging and nesting sites year after year and groups of families hatch at the same time and travel as a unit across the tundra between lakes to localised feeding areas (Prop *et al.* 1984, Loonen *et al.* 1998, Tombre *et al.* 1998). Because of their repeated use of sites it is easy to imagine that members of small flocks or colonies become familiar over time. These birds may even be capable of tracking the performance of neighbours through the years and using the information to choose among replacement mates. Regardless of the reasons, older birds apparently prefer other older birds and in some cases familiar older birds as replacement mates.

Eighty-two percent of the 946 partnerships terminated because of the death of one or both partners, 12% were divorces (when both partners were still alive but no longer together), and in 6% the situation for the pair break-up was not clear. An average of 17% of partnerships terminated annually due to the death of one partner. In terms of the whole population, this means that on average about 14% of all birds were unpaired each year and probably looking for a replacement partner. Based on the age structure of the population and annual proportion of yearlings, an additional 16% of the population were unpaired two- to three-year-olds that were potentially looking for their first mate. In spite of the mix of available ages, evidence suggests that the majority of geese tended to choose partners that were born in the same year.

Reproduction in relation to partners' age

Especially in long-term monogamous species like geese, the reproductive payoff of particular combinations of male–female attributes has presumably shaped the evolution of mating preferences. The choice of partner may depend on the state or characteristics of the chooser. For example, a young bird may prefer an older, more experienced mate but an older mate may not prefer the younger (Forslund & Larsson 1991).

To investigate whether the sexes contributed differently to the reproductive success of the pair, we divided the Svalbard birds into three age-classes (young 2–6 years, middle-age 7–11 years and old 12+ years), and plotted average brood success (presence of a brood in winter) for each male + female age combination (Figure 4.9). We found that pairs achieved the highest measure of reproductive success when both partners were in their prime, middle-age years.[15] Young females and old males had consistently low brood success, irrespective of their mate's age, indicating that they limited the success of the pair, whereas young males and old females could improve their reproductive success by pairing with middle-aged mates. This suggests that many birds should be attempting to gain middle-aged mates to improve their own reproductive performance. Since low breeding performance is attributable to young females and old males, the benefits of divorce and re-pairing could differ for the sexes at

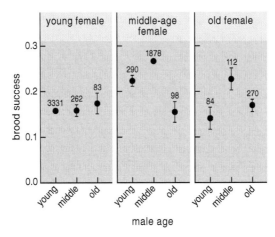

Figure 4.9. *Probability of returning to the wintering grounds with at least one gosling in relation to partner ages (i.e. brood success refers to proportion with broods in wintering area; sample sizes above SE bars). Age was partitioned into three age-classes: young and improving years (ages 2–6), prime, middle-age years (ages 7–11) and declining, old birds (ages 12+). From Black & Owen (1995).*

different stages of life. On the other hand, middle-aged individuals should avoid pairing with younger or older birds, because this would reduce their own reproductive success. Most geese would eventually achieve the best mate option (a middle-aged partner) by employing the strategy of choosing a similar aged mate at an early age and remaining faithful to that mate at least until the prime middle years. Whether individuals are able to achieve their preferred mate options is likely to depend on their own age or status and the availability of preferred mates at the time of pairing. In the next chapter we review the rare occurrences of divorce of long-term mates and discuss its consequences.

Reproduction in relation to partners' size

In both sexes, larger geese produce more offspring than smaller geese (Choudhury *et al.* 1996). At first glance this would suggest that all birds should attempt to pair with large mates. Indeed, the most successful pairs overall were large male–large female pairs and large male with medium female pairs (Figure 4.10). Interestingly, most geese maximised their reproductive performance by pairing with similar sized mates.[16] Small males and females did progressively worse with larger mates, whilst large males and females increased their breeding success increasingly with the larger mates. In other words, the larger the size mismatch the lower the reproductive success. This shows that reproductive success of a pair may be affected not only by their qualities as individuals, but also by their degree of compatibility or complementarity with a mate, at least in terms of size.

This compatibility idea was originally proposed by Coulson (1972), who studied Black-legged Kittiwakes *Rissa tridactyla*. He suggested that the compatibility of partners might affect the fitness of both individuals. The reason for the lower fitness in goose partners disparate in body size may be related to intra-pair aggression and lack of coordination. In geese, the male is larger than the female, and also more aggressive. Increasing size disparity of

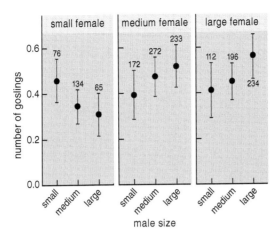

Figure 4.10. *Reproductive success (number of goslings in winter; sample size and SE bars) of different pair-size combinations. Barnacle Geese maximised production of goslings when paired to a mate of similar relative body size. Size categories were based on the first principal component (PC1) from skull and tarsus measures. Medium-sized birds were within half a SD of the average, small birds below and large birds above this. From Choudhury* et al. *(1996).*

mates could increase stress and risk of injury to females during social display and copulation, which can be quite aggressive. During copulations, for example, which occur in water, the male climbs onto the female's back and grasps the back of her head. Size disparity between the partners may influence how far the female is submerged under water, and how physically stressful and accurate the copulation process is. Small females may be overly stressed by social interactions with very large mates, thus inhibiting the reproductive process. In similar sized pairings, males may harass females less, thus enhancing the potential for coordination of duties, such as vigilance, defence of space and foraging routines.

Besides complementary body sizes and ages, birds may also prefer mates with compatible behavioural attributes or personalities (i.e. temperaments or coping styles; *sensu* Dingemanse *et al.* 2004, Spoon *et al.* 2004). In Greylag Geese, the correlation between male and female testosterone profiles was highest in the most successfully breeding pairs (Hirschenhauser *et al.* 1999). In Steller's Jays *Cyanocitta stelleri*, long-term partnerships comprising bold-bold and shy-shy ranked pair members nested earlier and fledged more young than pairs with mixed personality types (Gabriel & Black 2012). By making use of new methodologies to quantify personalities (Kralj-Fišer *et al.* 2007, 2010, Kurvers *et al.* 2011, 2012) future studies may more accurately describe the mate choice process and its consequences in geese and other highly social species. In the next chapter we ask whether all pair members maintain close contact with one another and explore the implications of long-term partnerships.

Summary and conclusions

Since geese compete for resources and care for young as a team and pair members often remain together for life, we might expect selection to favour the careful choice of partners. Pair formation consisted of a series of stages in social display, starting with the male directing excited calls and neck stretches at prospective females, which gradually increased in intensity when the female participated in the displays. The pairing process required mutual cooperation of both partners in social display, and the male's persistence in courtship was influenced by the female's responses. It seems that both sexes are making assessments about the alternatives.

An experiment lasting two years with 78 captive geese was conducted to identify mate sampling strategies. Females that were more vigilant after the appearance of a predator, and those that were larger and fatter, had the most trial partners prior to their final partnerships (maximum of six trial liaisons). Males may prefer these traits because reproductive success is highest for females that are capable of carrying more nutrient stores with them to the breeding grounds. Females with the darkest face patches also had many trial partners. In a second experiment we investigated whether geese choose mates because of their familiarity with one another or by choosing a population-specific phenotypic trait. They did not appear to discriminate between close and distant populations purely by phenotype. Instead, the geese preferentially sampled and established partnerships with individuals they became familiar with in the first year of life. In addition to familiarity with each other, selection may favour partnerships where both members are already familiar with the intricacies of a site, its terrain and the location of food and predators.

In the Svalbard population most pairs consisted of similar aged partners, even in replacement mates, and many pair members had grown up in the same Arctic neighbourhoods. Reproductive success was highest for pair members that were middle-aged. Low reproductive success in the early and later years was attributable to young females and old males, respectively. Relative sizes of pair members also affected reproductive performance. Most birds were able to maximise their reproductive success by pairing with relatively similar sized partners; the larger the size disparity of mates, the lower the breeding performance. This supports the idea that compatibility of mates may be important in determining the fitness of a pair.

Statistical analyses

[3] There was no significant difference in age of first pairing between the sexes (χ^2 = 2.64, df = 3, P = 0.50). Comparison of age of first pairing (one, two, three or four years of age) for 262 females and 192 males. The analysis was limited to pairs where both partners were ringed. Data from all years were combined since there was no significant change in age of first pairing during the study. From Choudhury et al. (1996).

[4] Frequency of initial pair bonds between seasons: females, 172 in summer and 90 in non-breeding season (χ^2 = 28.23, df = 1, P < 0.001). The same relationship was found for males: 123 in summer and 69 in non-breeding season (χ^2 = 15.19, df = 1, P < 0.001). There was no difference in this seasonal comparison between the sexes (χ^2 = 0.103, df = 1, P = 0.75).

[5] Length of trial liaisons and number of days away from nest initiation (r_s = 0.848, n = 26, P < 0.001). From Choudhury & Black (1993).

[6] Relationship between date of starting mate searching (February–June) and the number of trial partners (0–6 partners; r_s = 0.44, n = 62, P < 0.01). From Choudhury & Black (1993).

[7] ANOVA of male vigilance after the appearance of a stuffed fox in relation to the number of mates a male was paired to during the breeding season ($F_{1,14}$ = 5.69, P < 0.05).

[8] Trial liaisons with familiar or unfamiliar partners (χ^2 = 18.69, df = 1, n = 39, P < 0.001). See Figure 4.7A. From Choudhury & Black (1993).

[9] Final pairings with familiar or unfamiliar partners (χ^2 = 8.05, df = 1, n = 21, P < 0.005). See Figure 4.7A. From Choudhury & Black (1994).

[10] Trial liaisons with geese from same or different populations (χ^2 = 0.7, df = 1, n = 22, NS). See Figure 4.7B. From Choudhury & Black (1994).

[11] Final pairings with geese from same or different populations (χ^2 = 0.29, df = 1, n = 14, NS). See Figure 4.7B. From Choudhury & Black (1994).

[12] Familiar birds tended to pair earlier (average 166, SE 50 days) than unfamiliar ones (average 188, SE 22 days; t = -1.59, df = 21, P = 0.064). From Choudhury & Black (1994).

[13] Male body size (r_s = -0.378, n = 23, P < 0.05) and dominance rank (r_s = 0.454, n = 26, P < 0.05) were significantly correlated with timing of acquiring final mates. Principal component analysis was used to combine skull and tarsus measures to give a single index of overall body size (PC1). Dominance ranks were assessed from aggressive interactions, displacements and greeting displays. From Choudhury & Black (1994).

[14] General linear model of pair member ages for (A) first-time pairings: both partners had no previous mates ($F_{1,111}$ = 1045.7, r^2 = 0.90, P < 0.001); (B) second plus pairings: at least one partner had one or more previous mates ($F_{1,37}$ = 8.93, r^2 = 0.19, P = 0.005); (C) second plus pairings: both partners had one or more previous mates ($F_{1,16}$ = 3.82, r^2 = 0.19, P = 0.068). See Figure 4.8. From Black & Owen (1995).

[15] Overall effect of combined ages was significant, controlling for year (probability of breeding $F_{4,6404}$ = 4.81, P < 0.001), where middle-aged birds reproduced best. See Figure 4.9. From Black & Owen (1995).

[16] Pair-size significantly influenced overall reproductive success, after controlling for year and age (χ^2 = 21.5, df = 8, P < 0.01). See Figure 4.10. From Black & Owen (1995).

CHAPTER 5

Long-term partnerships

Wild geese are one of the few types of animals that practice the mating system referred to as perennial monogamy. Even within the bird world keeping the same partner year after year is not very common, though it does occur in 21% of the 159 avian families (Black 1996). However, in many of these species partnerships are only part-time, where pair members reunite only for the breeding season (e.g. Bried *et al.* 2003). In geese, monogamy is taken to the extreme because once initial mates are chosen many pairs stay together each day, each season, and often for life. In this chapter we ask whether persistence in pair bonds results in the production of more offspring, what mechanisms are behind such improved performance, why it pays to stay with just one mate for life and under what unusual conditions divorce occurs.

The pair bond

At each sighting of a ringed bird at the wintering grounds in Scotland, we routinely checked adjacent birds for additional leg-rings and recorded the identity of the mate and offspring. A number of behavioural cues can be used to determine the identity of pairs, including synchronised behaviours in (i) vigilance, (ii) feeding, (iii) defence of foraging space, (iv) common travel paths taken through the flock, (v) maintenance of proximity while foraging and in flight, (vi) coordination in social display and (vii) preflight signalling. The date of pair formation with a particular partner was taken as the first of multiple sightings when the pair was recorded together. The end of their partnership was attributed to the last date they were recorded together.

Monogamous partnerships were by far the most common mating strategy in Barnacle Geese: of 6,297 pair-years, 99.6% were one-to-one female–male partnerships (Black *et al.* 1996). We recorded only 28 cases when a third party joined an existing pair for between 10 months and four years. Barnacle Geese live on average for 9.5 years (maximum 27 years), yet most geese had only one mate during their lifetime. Of 2,618 recorded lifetimes, 65% had one mate, 24.5% two, 7.5% three, and 3% four mates or more (average 1.5; maximum seven mates). Pair bonds persisted for 1–16 years, with an average pair duration of 3.7 years. Average pair duration decreased with each subsequent mate, from 4.6 years with the first mate, 3.2 years with the second, 2.4 years with the third and 1.4 years with subsequent mates.[17] Most pair bonds lasted long enough that they ended because of death of one or both pair members (82% of 946 marked pairs).

Keeping in touch

It must be a challenge for Barnacle Goose family members to keep track of one another among masses of conspecifics, especially when a Peregrine attacks suddenly and the flock explodes in a panic. During the course of a typical winter day flocks take flight several times, resulting in an increasingly scattered distribution among marshes and pastures. Each time a flock takes off there is potential for pair members to lose track of one another. When flocks settle we inevitably see individuals walking erect and calling loudly until they are reunited with partners and family members. Sometimes pair members are separated for days or weeks. Ring-readers frequently made note of the unusual erect posture and persistent calling of a searching goose. This extra information told us that the bird was likely to be paired but temporarily separated from its long-term partner; an assumption that could be confirmed with examination of the long-term data. On examining the records in which searching behaviour was noted, males were found to have performed the behaviour more than twice as often as females: 567 and 231 records of searching by males and females, respectively. In addition to this conspicuous searching behaviour, biologists proposed three ways in which goose partnerships are maintained over time. All of the ideas are in need of further testing.

PREFLIGHT BEHAVIOUR Waterfowl usually perform a series of head and neck movements, referred to as head tossing or preflight signals, prior to taking flight. A context-specific vocalisation, which is a soft repetitive call, accompanies each head movement. These are thought to signal flight intention and coordinate take-off between partners (Raveling 1969a, Black & Barrow 1985, Black 1988). Based on data from 72 continuous focal samples when an observation ended in flight, males performed significantly more preflight movements than females (average 5.2 and 1.6, respectively)[18] and took flight first in 74% of the cases.[19] This indicates that males make a substantial investment prior to flight and that males generally lead and females follow in flights during the non-breeding season (Rees 1987, Black *et al.* 1996). In most cases the male delayed his take-off and continued to give head tosses until the mate looked up and also performed one or two preflight signals. Presumably, pairs that invest in pre-locomotory movements are less likely to become separated.

INCIDENCE OF TRIUMPH CEREMONIES Triumph Ceremonies consist of head tossing, neck stretching, wing waving, and a vocal duet performed by both members of the pair (review Figure 4.1). The displays are thought to reinforce or maintain the pair bond as well as signal the existence of the partnership and aggressive intentions to flock and colony members (Fischer 1965, Lorenz 1966, Radesäter 1974). In spring, prior to migration, noisy ceremonies frequently occur in foraging flocks, where one pair's display ignites additional displays among neighbouring pairs. When day length increased in March and the birds have more time to pursue other activities besides grazing, we recorded four Triumph Ceremonies per 1,000 birds per minute, an increase from the rate of one Triumph Ceremony per 1,000 birds per minute in the autumn and winter months (Black & Owen 1988). The fact that the rate of aggressive encounters also peaked in spring supports the idea that Triumph Ceremonies often precede and follow disputes among flock members (Black & Owen 1988). If the displays do reinforce the pair bond, pair members that frequently perform Triumph Ceremonies should be less likely to become separated during the day because the display results in greater proximity of partners.

Use of common roost sites Raveling (1979) suggested that pair and family members that become separated during the day may reunite at night-time roosts. Pair members that return to common sites, whether at roosts or foraging patches, would be more likely to find each other after temporary separations. This idea suggests a strong link between mate fidelity and site fidelity, which is a popular explanation for the maintenance of the monogamous mating system in birds (Ens *et al.* 1996, Cézilly *et al.* 2000). It was widely assumed that all goose species maintain constant contact with their mate and offspring (reviewed by Owen 1980a) until Johnson & Raveling (1988) argued that the smallest goose taxa probably evolved a more loose association with mates and family members than larger bodied geese because of the nature of their food and their need to maintain dense foraging flocks. In Chapter 6 we show that Barnacle Goose parents and goslings vary in the amount of time they spend together – that there is great variation in social relationships even within a single population. By scrutinising our database of resightings of marked pairs it became obvious that some partners were almost always seen together and others were not. In the next section we explore the variation in pair members' ability to maintain contact with one another during the non-breeding season, showing yet another way in which individual behavioural strategies vary in Barnacle Goose society.

Pair bond tenacity during the non-breeding season

With repeated resightings it is possible to build a picture of how frequently birds are with their long-term partners, with other birds or alone. *Pair bond tenacity* is an index of pair members' degree of association, namely the proportion of observations in which pair members are recorded as 'bird and mate'. We assumed that pairs that spent more time right next to each other would be more likely to be recorded together. We found that some pair members were only infrequently seen together from one day to the next while others were recorded together each and every time; annual pair bond tenacity values ranged from 22 to 100%, with an average of 71%. Figure 5.1 shows tenacity values for 230 pairs during the autumn and spring seasons. We found that pair members increased the amount of time together as the breeding season approached. Pairs were observed together significantly more often in spring than in autumn.[20] In the spring, pairs also perform many more social displays and males routinely defend feeding space for females (Akesson & Raveling 1982,

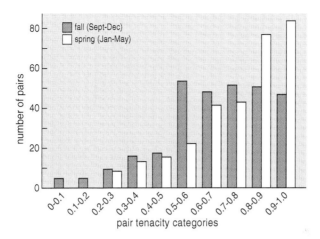

Figure 5.1. *Frequency distribution of pair bond tenacity in the first and second half of the non-breeding season (mid-point 1 January). Pair tenacity is the proportion of observations in which long-term mates were recorded together by observers. Pair members were seen together more often in the spring than in the autumn months. In spring pair members were recorded together for an average of 74.7% of observations (SE 1%), compared to 66.6% (SE 1%) in the autumn. Pair members were seen 10 or more times in both periods.*

Black & Owen 1988, Lamprecht 1989). This change in behaviour makes good sense if male assistance enables females to acquire fat stores in spring that will enable reproductive attempts, which is our main argument for why long-term pair bonds have evolved and are maintained in goose societies (Black *et al*. 1996). Male Greylag Geese ranked high in aggression, boldness and vigilance were also close in proximity to their mates (Kralj-Fišer *et al*. 2010). Detailed studies of Brent Geese have shown that females that were paired to more attentive males, with respect to chasing off nearby flock members in spring, enjoyed improved foraging opportunities and were more likely to breed the next summer (Teunissen *et al*. 1985).

When pair members were not with their long-term mates they were recorded with an unringed bird, another ringed bird, or they were unpaired (Figure 5.2). Typically, we spent about three minutes on each observation and in some situations, when the geese were feeding in dense flocks or in tall grass, it was not possible to check all legs for rings. In those situations the mate status column on the recording sheet was left blank. Pair members were recorded as unpaired only occasionally (2.2%). This status was attributed to occasions where birds were obviously not associating with any of their nearest neighbours. The conspicuous mate searching behaviour described above was our clearest indication that a bird was not with its long-term partner. Some of the records where pair members were recorded with marked birds other than their long-term partner were probably made in error since the primary goal of ring-readers was to work quickly and record as many rings as possible. However, it was not uncommon for mates to be recorded with a particular 'other' bird on more than one occasion, and in a few cases future mates could be traced back to these early encounters. In some instances a long-term mate was late in returning and the partner would begin associating with a new bird. For example, a male bird, XJV, was seen with PUX on 11 occasions in the first months of the wintering period in 1989. Apparently, the original

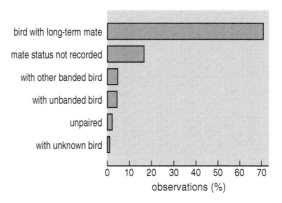

Figure 5.2. *Distribution of mate status records for 230 pairs where both partners had alphanumeric leg-rings. Based on 9,312 observations during 300 pair-years amounting to an average of 31 observations per annum per pair. In the majority of observations pair members were recorded together, indicating the nature of the pair bond in geese.*

mate, PNF, was late in returning on migration; the original partnership (XJV and PNF) was maintained over the previous two years. When PNF finally returned in December, the original pair was immediately re-established and was recorded together on 110 more occasions over six subsequent years.

We were interested to find out whether tenacity of pair members would increase or decrease with pair bond duration. We could have hypothesised that pair members would be seen together more often with each additional year of the pair bond if the pair was fine-tuning their cooperative behavioural repertoire in food finding, fighting for foraging space and sharing vigilance routines (*sensu* Black 1996). However, we could also hypothesise that pair members would be able to decrease the amount of time with a partner once their 'teamwork' had become established and they were confident in their mate's abilities. We limited this analysis to pairs known to have been together for at least one year and excluded pairs in their last year together, since some pair members were apparently not together as often in initial or terminal years.[21] The analysis, therefore, included 115 established pairs that were seen together for a minimum of 20 times in the seven-month non-breeding season and 10 times each in the autumn and spring seasons.

Figure 5.3 shows a plot of the negative relationship between pair tenacity and pair bond duration, where pairs were less likely to be recorded together as pair duration increased (i.e. with age of the pair).[22] Average tenacity values ranged from 79% for pairs that had been together for one year to 55% for pairs that were together for five or more years. The negative trend between pair tenacity and pair duration supports the second hypothesis outlined above. Pair tenacity may reflect an increasing familiarity between long-term pair members that no longer need to maintain continual, side-by-side contact. This finding is particularly interesting because we will show in the next section that a pair's reproductive success improves with increasing years with the same partner. It follows, therefore, that the measure of pair tenacity assigned during the non-breeding season was not positively related to the pair's reproductive success in previous years[23] or the next year.[24] Pair members seem to be able to gain benefits of long-term partnerships in terms of enhanced reproduction while reducing their investment in maintaining continuous, side-by-side contact with their

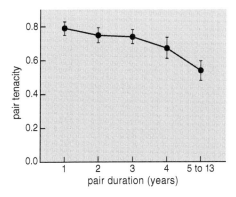

Figure 5.3. *Relationship between pair tenacity and pair duration for 115 pairs. Pair tenacity is the proportion of observations in which long-term pair members were recorded together through the wintering period. To ensure continuity in the dataset all pairs were together for more than 12 months and were not in their last year together. Each pair was observed on more than 20 occasions during the non-breeding months (Sept to May). There was a significant negative correlation between pair tenacity and pair duration, where older pairs were recorded together on fewer occasions.*

partner. Perhaps long-term partners are better at keeping track of one another over greater distances by using louder calls than young pairs (see below). And perhaps long-term partners are better at finding each other after temporary separation by returning to common foraging and roost sites (i.e. site fidelity) – an idea about monogamous mating systems that remains untested.

The fact that some pairs were consistently more likely to be seen together than others may indicate the existence of a gradient in birds' ability to maintain proximity to partners. In an intriguing experiment with 18 human-reared goslings, Lamprecht (1984) quantified individual differences in the birds' 'internal divorce tendency', meaning strength or degree of attachment to a human foster parent. By counting their vocalisations he could predict the degree to which the goslings would stray from the parent when given an option to follow a stranger or semi-stranger. All things being equal, such findings with young goslings may reflect their innate tendency to stay with future partners.

Long-term pair bonds: monogamy in the extreme

We tested in two ways whether there was a reproductive advantage to remaining with the same partner in multiple years. We first analysed if the number of years that partners had been together was related to the number of mature goslings that was produced (counted on the wintering grounds), while controlling for variation due to the year of the study and age of the birds. We found that the annual average production of offspring increased with the duration of the pair bond up to the seventh year and then decreased in later years (Figure 5.4).[25] This pattern suggests that middle-age pairs outcompete younger and older pairs for resources that enable successful reproduction. There was also a significant interaction between year and pair duration, suggesting that long-term association with one partner was more beneficial in some years than others.

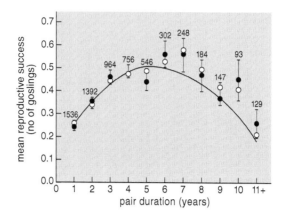

Figure 5.4. *Relationship between pair duration and reproductive success (goslings returning with parents to the wintering grounds; n = 6,297 pair-years). Filled symbols represent observed averages and their SE. Open symbols represent fitted values of log-linear models after controlling for year and female age. Sample sizes indicated. Reproductive success varied significantly with pair duration. From Black et al. (1996).*

In a second analysis we limited the data set to a single cohort (hatched in 1976) that lived at one colony (Nordenskiöldkysten), thus removing potential bias due to individuals experiencing different access to resources. We also used lifetime reproductive success (total number of offspring produced in a bird's lifetime) as a measure of reproduction for each bird. In this sample the average pair bond duration was five years (n = 119, SE 0.4 years), though 17 pairs were together for more than 10 years and three were together for 19 years. We checked reproductive success records for these pairs to determine whether longer-lasting pair bonds yielded more offspring. After controlling for lifespan and time unpaired, two factors that also influenced lifetime reproduction, more goslings (surviving to reach the wintering grounds) were produced by pairs that were together for more years (Figure 5.5).[26]

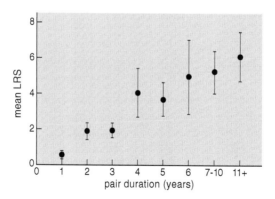

Figure 5.5. *Total number of goslings returning with parents to the wintering grounds (lifetime reproductive success, LRS) in relation to pair bond duration. Error bars indicate SE. Sample sizes: 19, 23, 21, 9, 13, 9, 14, 11. From Black (2001).*

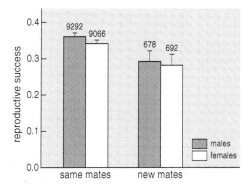

Figure 5.6. *Observed number of goslings returning with parents to the wintering grounds (i.e. reproductive success) in the first year after acquiring a new mate compared to pairs that were still intact, i.e. with the same mates. Males and females that changed mates returned with fewer young in the first year with their new mates. SE and sample sizes indicated. From Black* et al. *(1996).*

Constraints on alternative mates

The higher reproductive success of pairs with long-lasting partnerships may explain why the mating system of long-term, continuous partnerships has evolved and why it is currently maintained in goose societies. However, the mating system might also have evolved because of the high costs of seeking alternative mates. In goose flocks where competitive ability and access to food and nesting territories is influenced by the presence of a mate, birds that take time off from continuous partnerships to seek a different mate seem to suffer a setback in terms of their lifetime reproductive potential (Owen *et al.* 1988, Forslund & Larsson 1991, Black 2001). It is clear that when a mate dies, the surviving individual generally suffers a reduction in offspring production in the next summer. Reproductive success was significantly lower in the first year after re-pairing (i.e. for pairs comprising new partners) than in pairs with established mates (Figure 5.6).[27] However, by the second year this initial disparity was no longer apparent. We suspect that pair members need some months to establish a finely tuned relationship that will result in the acquisition of adequate resources, including food and nest sites. The penalty disappears in the second year with the new partner. Martin *et al.*'s (1985) male removal experiment with Lesser Snow Geese in Canada was particularly revealing. Females widowed just before egg-laying were unable to establish or maintain territories within the colony, as aggressive neighbours would physically attack and attempt to copulate with them. Similarly, Nicolai *et al.*'s (2012) experimental break-up of Black Brant *Branta bernicla nigricans* pair bonds in Alaska resulted in a reduction in female survival rate from 0.85 to 0.69, indicating a substantial cost to mate loss.

Teamwork and the competitive edge

Long-term partnerships may be selected for in goose societies because of the constant need for female–male cooperation throughout the annual cycle, including long-distance migration and short Arctic summers. Male assistance is apparently essential for females to acquire enough fat and nutrient stores to enable breeding attempts. In winter and spring, males act as sentinels and fend off competitors while females spend most of their time feeding (Black & Owen 1989a,b). In summer, the pair acts together in fighting for and maintaining a territory within the colony (Inglis 1977). In Barnacle Geese, males stand guard and defend eggs from patrolling gulls while females take short incubation breaks away from nesting territories (Prop *et al.* 1984). Each of these behaviours may be candidate mechanisms behind the improvement in reproduction with increasing pair duration. Individuals of each sex may invest more in their particular role each year pair members are together. The next two subsections outline some specific behavioural changes that correlate with pair durations in Barnacle Geese.

FLOCK POSITION In Chapter 8 we argue that despite the higher energetic costs of increased vigilance and conflicts with neighbours, dominant geese value positioning themselves at the edge of the flock because they acquire the first bite of the vegetation (see also Teunissen *et al.* 1985, Prop and Loonen 1989, Black *et al.* 1992, Rowcliffe *et al.* 2004). One of the suspected benefits of a long-term association is enhanced teamwork to monopolise prime feeding areas. To test these ideas, we checked whether the length of association with a mate was effective in describing the tendency to forage on the edge rather than the middle of flocks. We chose 270 pairs (without offspring) that were seen eight or more times in a given winter season. Flock position was significantly related to pair duration, with longer-term pairs spending more time in edge positions than shorter-term pairs.[28]

VOCAL CONTACT A study conducted on the semi-captive flock of Barnacle Geese at Slimbridge showed that the proportion of different types of calls given by a pair changed with increasing pair duration.[29] As pair duration increased, the proportion of soft calls declined, medium intensity calls increased slightly and louder calls increased the most. Researchers have yet to investigate the role of vocalisations in wild goose societies so we can only guess why the proportions of call-types change as partnerships grow older. The soft 'contact calls' are thought to help maintain contact between partners at close range, while the louder calls often precede aggressive interactions and maintain long-range contact (Collias & Jahn 1959). This shift in call type may enable older pairs to keep track of one another at greater distances while in large flocks.

Constant chatter is one of the most notable aspects of goose watching, yet this is one of the least-studied phenomena (Collias & Jahn 1959, Hausberger & Black 1990, Hausberger *et al.* 1991, 1994, Whitford 1998). Identifying the form, context and function of goose vocalisations in terms of intention of signallers and perception of receivers and eavesdroppers (*sensu* Johnstone 1997) would go a long way toward bettering our understanding of social complexities in goose societies.

Perennial monogamy in perspective

A multi-species comparison of mate fidelity (and divorce) in birds by Ens *et al.* (1996) supports the notion of a strong link between continuous partnerships and site fidelity. In the review of over 100 different bird populations of 76 species, mate fidelity was highest (and divorce lowest) in species that were resident with continuous partnerships (pair bonds persisting throughout the year), and mate fidelity was lowest (and divorce highest) in species that were migratory with part-time partnerships (pair bonds that re-establish only for the breeding season). Especially for species with long-term, continuous pair bonds with a high degree of site fidelity, a well-established partnership may outweigh the potential benefits of divorce and starting over again.

In a wild goose system individuals strive to achieve a threshold in fat/nutrient condition in each season (Drent & Prins 1987). Failure to achieve one threshold will preclude the next, such that an appropriate breeding condition may depend on an individual's foraging performance throughout the year (Chapter 11). Under this assumption, females should encourage year-round pair bonds, because by teaming up with a partner females increase their dominance status and are able to devote more time to foraging, use higher-quality feeding areas and attain larger body reserves. These ideas are consistent with those developed for swans, which also maintain long-term partnerships (Scott 1980a,b, Rees *et al.* 1996, Kraaijeveld & Mulder 2002). However, this does not nullify the possibility that individuals of either sex can improve on their choice of mate should a better alternative become available. To better understand why mate fidelity is the norm, in the next section we explore correlates and consequences of divorce. Divorce refers to surviving pair members who are no longer together in subsequent breeding seasons.

Divorce in geese

Divorce was very rare in our study populations. In the Svalbard population there were only 118 cases in 1,536 well-established pairs. In this section we consider why some geese choose the rare strategy of changing long-term partners. Ens *et al.* (1993) suggested that divorce should occur when a better option is available for one partner. The probability of divorce, therefore, should increase when there is greater availability of alternative partners in the population. In geese, alternative mates become available from two sources, (i) when older birds are looking to replace previous mates that die (i.e. widowers) and (ii) when there is an influx of two-year-olds looking for their first mates. The first group of alternative partners may offer more opportunity to improve on existing mates because they are composed of 'experienced' birds. To test this idea we compared the annual divorce rate among years with different numbers and types of single birds in the population. We found that the divorce rate was highest in years when potentially more 'experienced' mates were available,[30] which supports the *better option hypothesis*. We also found that divorce was even rarer in years when there were numerous young, unpaired birds in the population,[31] suggesting that geese may be less willing to leave existing partners when the chance of re-pairing with an inexperienced bird is high.

Coulson (1966, 1972) suggested that divorce should be more prevalent in pairs that fail to breed than in those that succeed, where breeding success is the product of compatible partnerships. For Barnacle Geese the incidence of divorce was not significantly different in pairs that failed to reproduce compared with pairs that succeeded; 21% of both faithful and divorced pairs produced at least one gosling in the previous year.[32] This result is not surprising since the vast majority (*c.* 80%) of breeding-age pairs fail to bring goslings to the wintering grounds each year, and yet the average annual divorce rate is only 2%. Presumably individuals would be more likely to switch mates when the disparity in pair members' attributes is great – when there is some incompatibility between partners. There were a few divorces in our study that may have been caused by the deterioration of a partner's quality due to an injury.

Owen *et al.* (1988) suggested that the chances of separation should be greater with increasing population size, because it would become more difficult to relocate a lost mate. However, we found that divorce did not change with increasing population size,[33] staying within the range of 0.5–5.4% annually. A review that compared population size and migratory distance of several goose species supports the conclusion that partners do not seem to have difficulty in keeping track of one another even in large flocks or when travelling large distances (Black *et al.* 1996).

So why is divorce so rare in goose populations? If the aim of divorce is to find a better, more complementary mate then we would anticipate a measurable improvement in reproductive performance with the new mate (Ens *et al.* 1996, Moody *et al.* 2005, García-Navas & Sans 2011). To detect whether there was an improvement after divorce, we compared the success of pairs in the year before and two years after 63 divorce events; the first year after was not considered due to the reduction in success after any kind of mate change (see above). Even after controlling for age in the analysis, reproductive success (i.e. the number of goslings brought to the wintering grounds) did not improve with the new mate after divorce events.[34]

Perhaps the risk of not finding a better mate and remaining unpaired for a time is enough to limit the incidence of divorce (Forslund & Larsson 1991, Choudhury 1995). Furthermore, the probability of divorce decreased significantly with increasing pair duration (Figure 5.7).[35] For example, 3–4% of partnerships lasting less than five years terminated in a divorce, compared to only 1% of those lasting eight or more years. This makes good sense because reproductive success increased with each year pair members were together. As pair bonds age and improvements are experienced, the pay-off to divorce and starting all over with a new partner will decrease.

In our study populations, the constant teamwork strategy currently yields the highest reproductive success. We could imagine that more frequent mate switching might be favoured where females are less constrained in their need for male help (*sensu* Gowaty 1996) or when

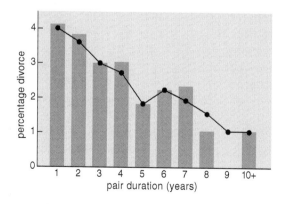

Figure 5.7. *Influence of pair duration on divorce. The probability of divorce was significantly related to pair duration. Filled symbols represent fitted values after controlling for year and age. Histograms are observed values. From Black* et al. (*1996*).

the sex ratio in the population becomes skewed, allowing one sex more opportunities to improve on initial mate choices. If geese experienced a large-scale change in sex ratio in a population (*sensu* Lehikoinen *et al.* 2008) the prevalence of strict social monogamy may be challenged and the incidence of divorce may rise. The best way to test many of the ideas regarding mate fidelity and divorce is with well-designed experiments that manipulate mate qualities or create particular alternative mate scenarios (Otter & Ratcliffe 1996, Cézilly *et al.* 2000, Heg *et al.* 2003, van de Pol *et al.* 2006b).

Summary and conclusions

In geese, monogamy is taken to the extreme because once initial mates are chosen many pairs stay together in close proximity, often for life. Most pair bonds lasted long enough that they ended because of the death of one or both pair members; 65% of individuals had just one mate and 25% two mates in their lifetimes. Ritualised searching behaviour and preflight signalling seemed to reduce the likelihood of pairs losing track of one another. Males displayed these behaviours more than females during the non-breeding season. Pair members participated in social displays (Triumph Ceremonies) signalling their partnership to each other and flock members. These displays intensified four months prior to the breeding season when females began to store fat and nutrients. Pair tenacity during the non-breeding season (i.e. the proportion of observations in which mates were recorded close together) varied from 22 to 100%. Pair members were more likely to be recorded close together in the spring than the autumn. Surprisingly, pair tenacity decreased with increasing years that pair members were together. Older pairs apparently did not need to maintain a constant nearest neighbour contact.

Annual reproductive success increased in the first seven years that pairs were together and declined in later years. After controlling for lifespan and time unpaired we showed that pairs that were together for more years produced more goslings. In goose society, where success in acquiring fat and nutrient stores in the non-breeding season influences the outcome of future breeding attempts, continuous teamwork seemed essential. Coordination of behavioural routines may improve with increasing years with the same partner. Whereas males benefit from access to a mate capable of reproducing on arrival in the Arctic (i.e. with sufficient fat and nutrient stores), females benefit from the male's role as protector, which enables access to food and nest sites. However, it is best to view pair bonds not as magical marriages but as flexible associations that can potentially end any time another option arises. Pair members may continually test the partnership to assess whether they should remain with the current mate. Divorce was a rare occurrence in Barnacle Geese, though there was some evidence that it occurred more in years when more single experienced potential mates were available in the population. Perhaps the risk of not finding a better mate and remaining unpaired for a time was enough to maintain the mate fidelity phenomenon.

Statistical analyses

[17] Birds had up to seven mates in a lifetime; most had just one or two. Pair bond duration (age of pair bond in years) was a function of mate number ($F_{3,2448} = 66.76$, $P < 0.001$), where the first partnership lasted the longest and the last one lasted the shortest period of time. No difference existed between the sexes in either number of mates or in pair duration. To reduce the possibility of accepting trial partners as true partners, for this analysis we excluded associations lasting less than two months, and only included pairs that spent at least one breeding season together (from Black et al. 1996).

[18] Males performed significantly more preflight movements than females in 72 observed flight episodes (Kruskal–Wallis test $\chi^2 = 33.5$, df $= 1$, $P < 0.001$).

[19] Two by two contingency table for males and females that either took flight first or second in 72 flight episodes ($\chi^2 = 14.5$, df $= 1$, $P < 0.001$).

[20] Pair members were seen together significantly more often in the spring (Jan–May) than in the autumn (Sept–Dec; paired t $= -6.77$, df $= 299$, $P < 0.0001$). See Figure 5.1.

[21] Pair tenacity was a function of pair status, comparing (i) those in their first year, (ii) those in their last year, and (iii) those in neither their first nor last year together ($F_{2,295} = 4.08$, $P = 0.019$). Average tenacity (% of observations when mates were recorded together) for the three categories were 71.3% SE 1.2%, 68.5% SE 1.2% and 74.9% SE 1.9%, respectively. The model included the variable 'pair' to control for multiple contributions to the data set by some pairs. The overall model explained 67% of the variation in tenacity values ($r^2 = 0.670$). Due to the difference among these pair status categories we limited subsequent analyses to pair-years in category (iii); pairs were together for at least one year and not in their last year together.

[22] Variation in pair tenacity was a function of pair duration ($F_{6,178} = 2.69$, $P = 0.022$). The model included the variable 'pair' to control for multiple contributions to the data set by some pairs. The overall model explained 84% of the variation in pair tenacity ($r^2 = 0.842$). Pair tenacity was significantly and negatively correlated with pair duration ($r_s = -1.0$, n $= 5$, $P < 0.01$). See Figure 5.3.

23 Pair tenacity was not a function of previous reproductive success of the pair (known to have bred successfully in the past or not) ($F_{1,183} = 1.02$, NS). The model included the variable 'pair' to control for multiple contributions to the data set by some pairs.

24 Pair tenacity in the non-breeding months did not influence subsequent reproductive success. Average pair tenacity did not vary for birds that did or did not return with goslings in the subsequent wintering season (returned with goslings, n = 42, average 71.7% SE 2.5%; did not return with goslings, n = 247, average 70.8% SE 1.1%) (t = –1.97, df = 287, P = 0.767). Similar, non-significant results were obtained for autumn and spring pair tenacity values and for a reduced data set where bird age was controlled.

25 Reproductive success varied significantly with pair duration (GLLM χ^2 = 49.0, df = 1, P < 0.001). The quadratic function of pair duration was also significant (i.e. $duration^2$ χ^2 = 39, df = 1, P < 0.001), as was the interaction between pair duration and year (χ^2 = 43, df = 16, P < 0.001), meaning mate familiarity may have been more useful in some years than in others. See Figure 5.4. From Black *et al.* (1996).

26 General linear model: Lifetime reproductive success for 119 pairs with known records was a function of pair bond duration ($F_{14,104}$ = 3.17, P < 0.001; Figure 5.5), after controlling for lifespan (2–21 years, average 9.3, SE 0.5) and time unpaired (1–7 years, average 3.5, SE 0.1). The order of importance of these variables was lifespan, pair duration, time unpaired and an interaction between pair duration and time unpaired. From Black (2001).

27 Geese that changed mates returned with fewer young in the first year with their new mates; tested after controlling for year, age, and pair duration effects (males GLLM χ^2 = 14.0, df = 1, P < 0.001; females χ^2 = 13.0; df = 1, P < 0.001; Figure 5.6). From Black *et al.* (1996).

28 Relationship between flock position and pair duration. The logistic regression model controlled for age and current number of goslings in winter; parents with the most goslings spend most time in edge positions in winter flocks (Black & Owen 1989a). Other NS variables were dropped from the final model; i.e. bird type, years of previous breeding experience, sex and skull size. The main effects accounted for a significant change in deviance (χ^2 = 118.7, df = 6, P < 0.001); age and pair duration interaction was also significant (χ^2 = 12.0, df = 1, P < 0.001). From Hammond (1990).

29 Relationship between call type and pair duration. Log linear model of six call types produced by 21 pairs varying between one and 14 years together (pair duration χ^2 = 23.4, df = 5, P < 0.001) (Bigot *et al.* 1995; unpubl. data).

30 Relationship between annual divorce rate and the number of older, more experienced potential partners that were unpaired and available in the population (r_s = 0.585, n = 15, P = 0.022). From Black *et al.* (1996).

31 Relationship between annual divorce rate and the number of young, inexperienced potential partners that were unpaired and available in the population (r_s = –0.495, n = 15, P < 0.05). From Black *et al.* (1996).

32 Incidence of divorce in years after reproductive failure and success. After controlling for year, age and pair duration effects there was no difference in probability of divorce for pairs that did and did not produce goslings surviving through autumn (logistic regression: male χ^2 = 1.0, df = 1, NS; females χ^2 = 0.6, df = 1, NS). From Black *et al.* (1996).

33 Relationship between annual divorce rate and population size, which increased over the study period (r_s = 0.004, n = 15, NS). From Black *et al.* (1996).

34 Comparison of reproductive success the year before and two years after a divorce event for 63 pairs. Reproductive success data was adjusted for annual and age variations by calculating the deviation from the annual average for each age class (GLLM: males χ^2 = 0.06, df = 1, NS; females χ^2 = 0.88, df = 1, NS). From Black *et al.* (1996).

35 The probability of divorce was significantly related to pair duration (logistic regression: n = 5,637 females, χ^2 = 19.4, df = 1, P < 0.001; n = 5,616 males, χ^2 = 24.8, df = 1, P < 0.001) after controlling for year and age. See Figure 5.7. From Black *et al.* (1996).

CHAPTER 6
Family life

When the geese arrived back in Scotland we began our seven-month campaign of reading rings as flocks landed on the saltmarsh and pastures near the hides and observation towers at WWT Caerlaverock. We checked each set of legs for colour-rings and searched adjacent geese to determine the size and composition of the marked birds' families. Years following summer expeditions to capture and ring entire families when they were flightless provided an opportunity to quantify how long goslings associated with their parents during the first year. By autumn goslings were nearly full size but they were quite thin and still had juvenile plumage patterns, as well as gosling-like vocalisations and mannerisms. The majority of individuals walked in pairs, so it was relatively easy with repeated observations to determine which of the ringed adults were accompanied by maturing goslings. While watching family members and recording their rings we noticed some goslings would come and go from the family unit. Sometimes they strayed for several minutes, travelling among flock members. On their return they would call and lower their heads in a submissive 'greeting' posture as parents and siblings responded. We extracted several variables from the records for this chapter, including 'brood size' (number of goslings in the family), 'length of association with parent' (number of days till goslings left and did not come back to the family), and 'degree of contact' (proportion of records in which goslings were recorded with parents while still part of the family unit). In this chapter we show that there was great variation in the amount of care Barnacle Goose parents afforded their goslings, since some goslings left the family unit early on while others remained for much of their first year. Some family groups were still together even on return to the breeding grounds when goslings were 11 months of age. Essential skills develop in the presence of experienced parents that lead goslings into life among competitors, around predators, and through foraging areas.

Goslings have two major dilemmas in the first year: (i) how to cope with siblings, and (ii) how long to stay with parents. Parents must also solve issues concerning how much care to give to goslings. If their investment is too costly in one particular year, it might limit parents' chances of subsequent survival and further reproduction. We consider these dilemmas through an examination of detailed observations of individually marked birds in the wild and even closer observations during experiments with geese that we raised in captivity. We address these issues first from the offspring's perspective and then the parent's.

Offspring perspective

Unlike many other birds, goslings are not fed by their parents and once they have a full set of feathers and can fly they should be able to withstand cold weather and escape most predators. So why do most goslings stay much longer with parents? Figure 6.1 provides evidence that life in the family actually pays in terms of body condition. By the 10th month of age, after the first leg of the northward migration in Norway, family goslings were in better condition than those that were fending for themselves.[36] Being part of a family enabled goslings to forage with fewer interruptions. For example, five-month-old family goslings enjoyed an average of 49 seconds of uninterrupted foraging compared to only 18 seconds per bout for single goslings (Black & Owen 1989a). This was because single goslings suffered an average of 16 attacks per hour from aggressive flock members compared to only two per hour for goslings that were still part of a family (Black & Owen 1989b). Feeding with fewer interruptions probably enables birds to find more profitable food items, resulting in better body condition. Based on these comparisons alone we are unsure whether single goslings lose condition after they leave the family or if parents favour larger goslings to begin with; a question to which we return later in this chapter. The next two sections develop two additional advantages of family life for goslings.

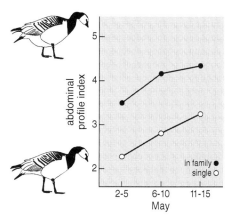

Figure 6.1. *Comparison of abdominal profiles for goslings in families and single goslings during three weeks on Norwegian staging islands. Family goslings deposited substantially more fat stores than single goslings. Average sample sizes are 24 (SD 6.8) for family goslings and 33 (SD 12.1) for single goslings. From Black & Owen (1989a). Each profile index value on the y-axis amounts to about 90g of body mass (Zillich & Black 2002, Madsen & Klaassen 2006).*

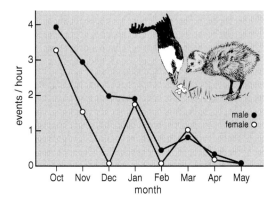

Figure 6.2. *Frequency of food sharing by female and male parents during autumn, winter and spring. Male parents shared food slightly more than female parents, males at 1.8 times/h (44 hrs of observation) and females 1.2 times/h (in 42.5 hrs), but this difference was not statistically significant. From Black & Owen (1989a).*

Learning to forage

Goslings are extremely inquisitive and will sample almost anything, including flowers, fruit and seeds, but also less nutritious plants like lichens. Our observations of foraging sessions suggest that goslings may learn from parents which food items to favour. Like other waterfowl, young goslings are equipped with a reserve of yolk from the egg that allows several days' time for exploring and sampling foods (Kear 1965). Within days goslings begin selecting items that are most digestible and high in protein (Sedinger & Raveling 1984). This initial food selection process is probably enhanced through episodes of food sharing with parents, when goslings feed from the plant on which a parent is feeding. Food sharing behaviour is most common in the brood-rearing areas, but it is also observed on the wintering grounds when goslings are 4–10 months of age. In October food sharing occurred at a rate of four events per hour, but steadily declined as goslings became older (Figure 6.2). This behaviour is analogous to Greater Snow Goose *Chen caerulescens atlantica* parents surrendering rhizome-digging holes to soliciting goslings (Turcotte & Bédard 1989).

Learning when to fight

Several of the chapters in this book describe the aggressive nature of geese and the foraging advantages that dominant birds enjoy. In this section we address the question of how aggressiveness develops. In particular, we asked whether goslings are equipped with an innate predisposition for aggressiveness and whether aggressiveness is learned by watching parents and fighting with siblings.

Our study of 23 captive goslings that were closely observed for their first 90 days suggests that acquisition of aggressiveness is indeed influenced by early experiences in the family. We hatched the goslings in an incubator and reared them in four groups (or broods) of siblings, but without parents. Once goslings dried off into tiny 'fluff balls' on hatching day they began pecking and tugging at things, including their younger siblings that hatched later in the day. Therefore, the initial pecking order within the brood was influenced by hatch order; the oldest birds became dominant over later hatched siblings even if the oldest was the smallest bird in the brood.[37] In some broods this age-mediated dominance rank was maintained for the first three months but it did not seem to influence dominance relationships when unfamiliar goslings met. Outside the family (without the influence of parents), unfamiliar goslings worked out dominance relationships according to relative body size and mass, where larger and heavier birds won more encounters.[38] Males were generally larger and heavier and were usually more successful in aggressive encounters than females.[39]

We designed two additional experiments to test the influence parents had on the development of gosling aggressiveness (Black and Owen 1987). First, we determined the aggressiveness of 54 pairs in the feral Barnacle Goose flock at WWT Slimbridge. Parental aggressiveness was measured by scoring nest defence behaviour while we approached to check eggs during the laying period. Nest defence was scored on a scale from one to five, where one refers to occasions when both parents stood their ground, hissed, bit and hit us with fighting bones normally concealed on the edge of their wings, and five refers to both members of the pair fleeing and staying well away from us. We also observed 28 aggressive encounters under natural conditions between the most and least defensive pairs. Twenty-one of 28 of these encounters were won by pairs that we ranked as very aggressive in defence of their nests. We assumed, therefore, that our measure of aggressiveness was also a good measure of the birds' aggressiveness among conspecifics.

We 'borrowed' 21 goslings (average age 75 days) from moulting flightless parents and introduced them to the 23 similar aged goslings that were reared in captivity without parents but among siblings. Both sets of goslings originated from eggs from the same set of semi-captive birds on the WWT Slimbridge grounds. We consecutively placed the two highest-ranking birds from each rearing-group in the same pen where a series of greeting postures, threats, pecks and chases ensued. A stable pecking order was eventually established in the four-bird groups, usually within 10 hours. In almost all the groups, parent-reared goslings were clearly more aggressive, and therefore dominant over sibling-reared goslings.[40] Goslings reared with parents were probably more adept at conflict assessment and fighting skills than those reared without parents. In a similar experiment with captive Hawaiian Geese, parent-reared goslings were dominant over others (Marshall & Black 1992).

We also kept track of the dominance relationships that developed among the 21 goslings that were 'borrowed' from Barnacle Goose parents. We housed these goslings in three groups according to their parents' aggressiveness and rearing situation: (i) goslings reared by the most aggressive parents, (ii) goslings reared by least aggressive parents, and (iii) goslings from least aggressive parents that were foster-reared by the most aggressive parents. Goslings in the last group hatched from eggs that were swapped between nests of parents at opposite ends of an aggressiveness rank order (Black & Owen 1987). Comparison of the goslings' success in encounters indicates that those raised with the most aggressive parents were more aggressive than the other goslings: they won 59% of 128 encounters. Goslings reared by subordinate parents won the fewest encounters (26%), while the foster-reared goslings scored intermediate to the others (45%). So it seems, as with many traits, that aggressiveness is shaped by early learning experiences within bounds inherited from parents (Halliday & Slater 1983).

The findings described in this section are in line with the idea that goslings reared with parents have more opportunities to learn appropriate behaviours that are useful in wild goose flocks. Learning which foods to eat and how to compete with conspecifics are essential skills. This leads us to question why not all goslings remain with their parents for as long as possible if the benefits are so great; a dilemma that will be dealt with in the following two sections.

Parent perspective

Parents, especially male parents, frequently direct threats or attacks at neighbouring flock members in order to maintain foraging opportunities for family members. Black & Owen (1989a) reported that parent males threatened or attacked other geese once every three minutes during a typical winter day in the pastures. This can be energetically demanding and potentially dangerous if conflicts escalate into grappling fights. Parents also devote much of their time to vigilant behaviour, being watchful for competitors and predators and monitoring family members. Parent males spent as much as 21% of their day being vigilant, compared to only 8% for paired males without goslings in attendance. Parent females were much more vigilant (head-up posture) than females without goslings (13% versus 4%; Black & Owen 1989a). By comparing time budgets of parents rearing small goslings, Pär Forslund (1993) showed that the intensity of vigilance behaviour was dependent on brood size because parents immediately decreased their vigilance postures when brood size decreased due to losses.

Devoting time to vigilance and encounters with neighbours, instead of near-constant foraging, may be costly for parents. December in Scotland only has seven hours of daylight and plant growth is negligible. At this time of year Barnacle Geese graze for more than 90% of the day and still lose body mass due to short days and cold weather (Owen *et al.* 1992). Limitations on foraging performance may be the reason why most goslings are expelled from families. Yet some parents endure the costs of parental care well into spring and during migration back to the breeding grounds. Is there some advantage to prolonged parental care?

We compared parents' future success in reproduction according to the length of their associations with offspring. Table 6.1 shows that parents that associated with their young through to spring (9+ months of investment) returned more often the next year with a new set of offspring than parents that split with their young earlier.[41] This seems to indicate

Table 6.1. *Length of association with offspring and the parents' success in bringing back at least one additional gosling to the wintering grounds in the following year. Parents that maintained contact with goslings well into the spring season were more likely to return the next year with more goslings, providing the basis for the gosling helper hypothesis. From Black & Owen (1989a).*

Timing of family duration (days)	Bred successfully two years in a row	Successful followed by an unsuccessful year
Autumn (92–154)	0	9
Winter (155–224)	0	19
Spring (> 224)	35	88

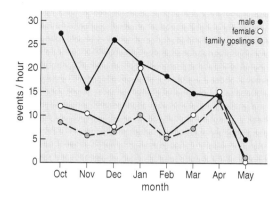

Figure 6.3. *Frequency of aggressive conflicts with neighbours by family members through the winter (parent males n = 679 events; parent females n = 404 events; family goslings n = 290 events). Older goslings and females increased their effort to match the male's in April. From Black & Owen (1989a).*

a clear advantage of maintaining contact with goslings into the spring season when food availability is no longer limited by short, cold days. By associating with at least one gosling, parents maintain the high social rank afforded to family units and gain access to prime feeding areas. Goslings that remained in the family into the spring season assisted parents in fighting for foraging space; their effort in competing with neighbouring flock members increased to match the parents' effort in April (Figure 6.3). Besides assistance in encounters, any contribution by goslings in detecting competitors and predators may enable parents to devote more time to foraging. The goslings' contribution to vigilance was never as much as their parents', but it did increase with age; from 3% of their time budget at five months of age to 6% at 10 months. This increase was the result of shortening the interval between head-up postures from once every 64 seconds at five months to every 49 seconds at 10 months (Black & Owen 1989a). We refer to this phenomenon as the *gosling helper effect*; i.e. when mature goslings increase their participation in the family tasks of maintaining foraging space and watching for competitors and predators. Under this scenario, parents that maintain their association with helpful goslings will have a competitive advantage.

A problem with concluding that the benefits of higher dominance status and gosling help better enable parents to achieve breeding condition has to do with a confounding variable, bird quality. It is possible that the parents in our analyses were good quality birds that would have bred successfully with or without maintaining prolonged contact with goslings. We address this possibility in a more detailed analysis presented in Chapter 9.

Family break-up

Due to the cumulative costs of parental care perhaps it would benefit parents to expel at least some of a current set of offspring as the next breeding season approaches. When goslings grow to full size and select plants at a rate of up to 200–300 bites per minute, parents may have to settle for second-best food plants, especially if several goslings remain in the family. A conflict between parents and offspring will appear when goslings begin to impinge on the

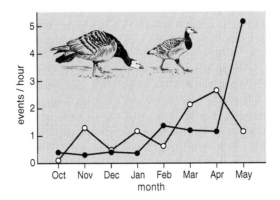

Figure 6.4. *Frequency of parental threats and attacks at their own offspring (n = 126 parents, black circles) and frequency of goslings' 'facing away' or greeting displays (n = 96 goslings, open circles). Gosling greeting postures seemed to appease parents' aggressive behaviour. Male parents were probably responsible for the eventual departure of goslings from family units. From Black & Owen (1989a).*

parents' ability to build fat and nutrient stores for the next migration and future breeding attempts (*sensu* Trivers 1972). At this point parents must decide which goslings to exclude from the family and which to allow to remain. In the next two sections we consider whether parents favour particular goslings in the process of family break-up.

Throughout the goslings' first year, parents direct an increasing number of aggressive threats, pecks and bites at their young (Figure 6.4). The only way goslings are able to withstand this is to employ a submissive 'greeting' behaviour that subdues the otherwise overly aggressive parents. The behaviour, which is also used by adult females to withstand advances by enthusiastic males, consists of a characteristic 'facing away' or greeting posture and noisy vocalisations (Radesäter 1974). Goslings use this behaviour when they approach or get approached by a parent or dominant sibling. Aggressive threats and attacks within families during 43 hours of observations of 135 families in winter were mainly directed by male parents at goslings (99 times observed), which was much more often than the attacks by females towards goslings (27 times) or between goslings and siblings (27 times). This suggests that male parents were largely responsible for the gradual break-up of families, but female parents and siblings also contributed. In the next section we consider some of the consequences for goslings staying with parents for differing amounts of time.

Parent–offspring association

We determined the length of association with parents for 543 individually marked goslings observed after they arrived in Scotland. The average date that goslings were last seen with parents (family departure) was early December, which was 70 days after arrival on the wintering grounds when goslings were 5.3 months of age (Figure 6.5). This analysis shows that about 30% of goslings that arrived on the wintering grounds were already separated from parents, another 30% became separated in the first 60 days in Scotland, and 18% remained in the family through to April/May and initiated northward migration with their parents. This distribution, however, varied considerably through the study period. Figure

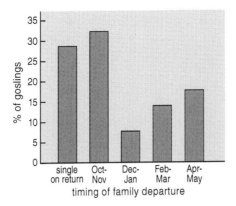

Figure 6.5. *Timing of family departure for 543 goslings based on dates last seen with parents. Goslings were observed in Scotland five or more times and at least once after 1 February (average of 11 resightings per gosling, maximum 40 resightings).*

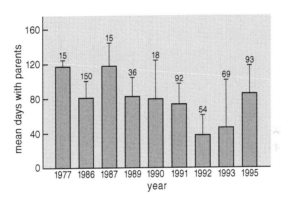

Figure 6.6. *Length of association with parents expressed in days on the wintering grounds from 1977 to 1995. Gosling associations with parents varied significantly among goslings' birth year. Average, SE and sample sizes indicated.*

6.6 shows the average length of association with parents expressed in days after arrival on the wintering grounds and how this changed from one year to the next.[42] Average annual length of association varied from a low of about 40 days in the winter of 1992 to a high of 120 days in 1977 and 1987.

The proportion of goslings returning to Scotland already separated from parents increased through the study with increasing population size.[43] This means that an increasing number of goslings did not enjoy the benefits of parental care in their first winter. Shorter lengths of associations might be due to the increasing difficulty of maintaining contact as the population size increased, with more birds, more noise and more mixing. Though logical, this idea was probably not fully responsible for shorter parent–offspring associations. We find it more likely that shorter lengths of associations were due to a decreasing benefit to maintaining the family. Full-grown offspring and parents may find it more profitable to separate when considering the trade-off between better foraging opportunities and time spent on vigilance and fighting with each other and with flock members.

Which goslings receive most care?

In this section we explore whether some goslings had access to more than their fair share of care from parents or if the amount of care was distributed equitably among siblings. This enquiry is often referred to as the dilemma of *sibling rivalry*, a topic that was reviewed by Mock & Parker (1997).

We proposed that those specific goslings which are allowed to remain in the family for the longest periods would be those that help with the family duties of maintaining foraging positions (and space) within the flock and watching for potential competitors and predators (Black & Owen 1989a). We have not been able to adequately test this 'helper' hypothesis in the wild. However, in the four families intensively studied by Stephanie Warren at WWT Slimbridge during a period of family break-up, goslings that were consistently closer to parents were attacked least by parents.[44] With regard to helping with vigilance duties, however, there was no relationship between closeness to parents and the amount of vigilance among goslings.

Figure 6.7 shows that the average length of association was shorter in larger families.[45] Although one or two goslings generally remained, most goslings from larger broods left the family soon after arrival in the wintering grounds. In other words, goslings reared with several siblings left the family much sooner than those from broods with just one gosling, suggesting that the presence of siblings influences the family departures. Furthermore, the length of association with parents was typically shorter for males than for females. Competition between male siblings was apparently more intense than between females or between males and females.[46] For example, for males in broods of two or more, the length of association was significantly shorter when with a male sibling (average 83 days, n = 81) than with a female sibling (average 122 days, n = 35).[47]

Based on observations of families that were watched at the staging area in Norway during their northerly migration (when goslings were 10 months of age) it was evident that parents were attacking specific offspring more than others and that a rank-order existed among siblings. When size among siblings was noted, parents attacked smaller goslings more than larger ones (13 out of 17 events), larger siblings also threatened smaller siblings (10 out of 10

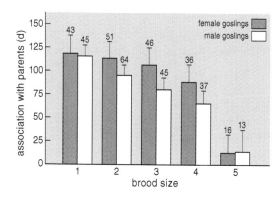

Figure 6.7. *Length of association with parents in goslings' first winter in relation to brood size. Goslings from larger broods were observed with parents for a shorter duration. In brood sizes 1–4, males were with parents less than females. Average, SE and sample sizes indicated.*

times) and larger siblings remained closer to parents in 12 out of 16 families.[48] In one family watched over a three-day period, two dominant siblings that were larger and fatter remained within a few metres of one of the parents and the smaller, thinner, subordinate sibling frequently strayed from the family to feed on its own up to 30 metres away. Upon return to the family, siblings and parents harassed the smaller gosling, such that if it remained nearby it had little time to feed.

In Canada Goose *B. c. moffitti* families in northern California, Ken Griggs found that parents maintained contact with male offspring longer than females and that these young males were more aggressive and vigilant than young females. Among female offspring, parents afforded more care to those that were most aggressive and vigilant. He also found that female parents were able to spend more of their day feeding in families containing more helpful offspring (Griggs 2003). These findings support the *gosling helper hypothesis*.

Gosling life without parents

Even though single birds are at the bottom of the pecking order in geese (Black & Owen 1989b), there must come a time when it would be better to be a single than to stay in the family and suffer threats and attacks from increasingly aggressive parents and dominant siblings. This departure time is likely to be governed by a gosling's ability to obtain a daily ration of food to maintain body condition (i.e. the Daily Existence Energy). Goslings have several alternative strategies available once they leave their parents: roam singly, form groups with other immatures, or follow unrelated family units, parasitising their vigilance and dominance status (Black & Owen 1984, Siriwardena & Black 1999). Maintaining contact with 'social allies' in Greylag Goose families has been shown to reduce levels of corticosterone (stress hormone), especially in females (Scheiber *et al.* 2009). The costs and benefits of these alternative strategies may affect the duration of a gosling's association with its parents. In the next two sections we show evidence suggesting that goslings remaining with their parents gain long-term benefits in terms of survival and reproduction.

Long-term benefits from association with parents

The survival of 625 individually marked goslings was positively influenced by several measures describing their early family life, including brood size, degree of contact with parents, and length of association with parents.[49] Gosling viability (survival from age 3–15 months) was highest for those reared in larger broods and those that spent most time in the family in the first year.

The next step is to determine whether the length of association with parents also positively influenced reproductive prospects. This analysis included records from the 1986 cohort, including 59 males and 61 females that bred successfully for the first time (if at all) between their 2nd and 10th years. These birds (when goslings) spent between 0 and 231 days with their parents on the wintering and spring staging areas (records began when birds arrived in Scotland, 27 Sept). To be included in this analysis they had to be observed a minimum of five times (average 11, max 40) spread over the first winter and spring. The analysis

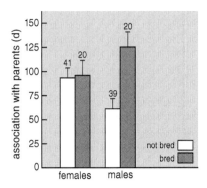

Figure 6.8. *Length of association with parents (days) in goslings' first winter and future reproductive status (bred successfully or not) in the first 10 years of life. Males that eventually bred associated with parents for longer in their 'gosling year' than those that did not breed. Successful breeding was determined on the wintering grounds. Average, SE and sample sizes indicated.*

showed that males that eventually bred successfully at least once had associated longer with their parents in their first year than those that did not breed (Figure 6.8).[50] There was no relationship for females.

Figure 6.9 shows the variation in 'degree of contact' with parents, determined by the proportion of observations in which goslings were seen with parents. The analysis was for 141 marked goslings with long associations with parents through the winter and into the spring period, with a minimum of five records. Even when the family duration lasted into spring, some goslings were rarely observed close to parents while others were almost always seen in the family unit. The degree of contact with parents positively influenced gosling survival (above), but did not influence subsequent reproductive prospects.[51] Apparently, during our study, it was enough for goslings to maintain some contact with parents, even periodic contact, to reap the benefits of prolonged association.

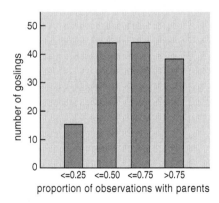

Figure 6.9. *Frequency of goslings' degree of contact with parents as measured from the proportion of observations with parents. Most goslings were seen with their parents for more than half of their observations. This index of 'togetherness' was not a function of birth year, brood size, family sex ratio, or gosling sex. The analysis included 141 goslings that were observed with parents through the winter and into the spring period, with a minimum of five observations.*

In a study of Canada Geese, Raveling *et al.* (2000) also showed that survival was better in goslings that associated with parents than in those that did not. In addition, they showed that goslings experiencing long-term stability in family associations, especially those in large families, were most successful at rearing a brood as two-year-olds. Much of the mortality in their study was caused by hunting; young birds are often more vulnerable to hunting, and probably more so when separated from parents. Our study of a fully protected population of Barnacle Geese suggests that non-family young were apparently also more vulnerable to 'natural' predators or other factors affecting mortality Furthermore, in Hawaiian Geese, goslings reared by parents responded more appropriately to a passing predator by remaining vigilant longer than those reared without parental contact (Marshall & Black 1992). Conditions and learning opportunities experienced in early life are of great importance in animal societies, not least in geese and other avian systems (de Kogel 1997, van de Pol *et al.* 2006a, Kriengwatana *et al.* 2013).

Unanswered questions

A suite of hypotheses is emerging suggesting that parental investment at the egg stage may affect actions and other phenotypic characteristics of offspring after they hatch (Dufty & Belthoff 2001, Stamps & Groothius 2010). For example, parent females may transfer 'information' about local conditions to offspring via hormones that are delivered to the eggs. Chicks may then develop adaptive responses to current conditions. Parent females may also transfer variable amounts of hormones into different eggs in the clutch, thus influencing developmental conditions for particular chicks (Schwabl 1993). It would be valuable to consider whether early hormonal conditions shape gosling explorative behaviour, aggressiveness and attachment strength, thus influencing how long goslings may stay in the family group. Waterfowl systems, and especially those of geese, would make ideal systems to test some of these ideas and issues regarding differential parental investment in eggs.

Summary and conclusions

Goslings enjoyed substantial benefits from family life: family goslings were chased less, had fewer interruptions while foraging, and achieved larger fat stores prior to migration than non-family young. Food sharing events, where parents gave up food items to their goslings, may help goslings learn which plants are most appropriate to eat. Goslings also acquired useful fighting assessment skills during their first year of life. Dominance relationships within broods were determined by the order of hatching, where younger birds submitted to older birds. Outside the family, however, conflicts among similar aged goslings were usually won by heavier and larger individuals. A series of detailed experiments showed that a gosling's aggressiveness was linked to rearing experiences and inherited parental tendencies.

Most goslings stayed with their parents for between four and 11 months. Goslings probably choose an appropriate time to leave the family, based on their ability to obtain enough food without being harassed by dominant siblings and aggressive parents. If faced with parental aggression, goslings adopted a strategic 'greeting' display that seems to subdue it. In February, when goslings were about eight months of age, parental attacks intensified and goslings' greetings diminished. The number of single goslings increased as parental aggression increased. Outside the family unit, goslings roamed wintering flocks as singles, with siblings or in other groups. Gosling size or sex did not influence length of association with parents. We hypothesised that parents should allow the most 'helpful' goslings to remain in the family for longest, i.e. those that help maintain foraging positions within the flock and scan the horizon for potential competitors and predators. Parents observed with mature goslings at 9–11 months of age were more likely to return with another set of goslings the next year. Goslings from large broods and those staying with parents for longer periods had a higher survival probability. Male goslings that associated with parents for longest were more likely to breed successfully. These relationships may explain why prolonged parental care has evolved and is maintained in wild goose societies.

Statistical analyses

[36] Family versus single goslings fatness scores were tested by categorising the number above, below and at the median value (n = 170). The test was highly significant (χ^2 = 40.6, df = 2, P < 0.001) in favour of fatter scores for family goslings. From Black & Owen (1989a).

[37] A stable linear rank order developed within broods after three weeks. This initial rank at 28 days was positively correlated with hatch order (age) in three of four broods (Spearman rank correlation P < 0.05) and with body mass at 21 days in two of four broods (P < 0.05), but not with body mass at hatch. Older birds tended to dominate younger ones when contestants' body masses were within 10 grams (P = 0.011, n = 10, sign test). From Black & Owen (1987).

[38] Social rank (pecking order) of 77-day-old goslings that had not previously met was positively correlated with age (Spearman rank correlation P < 0.01), body mass (P < 0.05) and body size (P < 0.001). Tests were among 27 unfamiliar goslings. From Black & Owen (1987).

[39] Males won more encounters than females (χ^2 = 42.7, df = 6, P < 0.01). After controlling for body size, males still won more than was expected by chance alone (P = 0.008, n = 14, sign test).

[40] Results of contingency tables (wins and losses) comprising parent-reared versus nonparent-reared goslings. Goslings were placed together based on their within-group ranks; e.g. high-ranked birds were placed together with other high-ranked birds. Parent-reared birds won more encounters than nonparent-reared: χ^2 = 11.5, df = 1, P < 0.001 for goslings with high ranks; χ^2 = 14.3, df = 1, P < 0.001 with intermediate ranks; χ^2 = 5.1, df = 1, P < 0.05 with low ranks. Goslings from both rearing styles were of similar body mass and size. From Black & Owen (1987).

[41] Two by two contingency table of breeding success comparing autumn + winter versus spring lengths of association for pairs with marked female parents (χ^2 = 10.37, df = 1, P = 0.001).

[42] Variation in length of association with parents was strongly influenced by gosling birth year ($F_{8,534}$ = 3.96, P = 0.0002; Figure 6.6); but not by gosling sex. The slight decreasing trend with year was not significant (r_s = –0.533, n = 9, P = 0.139). See Appendix 1, Table 1 for full details of analysis concerning variation in length of association during goslings' first year in relation to birth year and gosling sex.

[43] Percentage of goslings (singles) arriving on the wintering area already separated from families in relation to population size (r_s = 0.729, n = 9, P = 0.026). There were more single goslings in autumn at higher population densities between 1977 and 1995.

[44] Spearman rank correlation for eight goslings in four families; testing gosling aggressiveness and threats/attacks by parents, r_s = 0.90, P < 0.01. These semi-wild (feral) families were watched for 10 minutes each day for eight weeks and the distance between each gosling and parent was recorded at one-minute intervals. The number of intra- and inter-family aggressive encounters was continuously recorded.

[45] Length of association (days) varied with brood size (females $F_{6,186}$ = 3.70, P < 0.002, males $F_{6,207}$ = 2.80, P = 0.012; Figure 6.7). See Appendix 1, Table 2 for full details of analysis testing the link between length of association and brood size, controlling for birth year.

[46] See Appendix 1, Table 3 for full details of General Linear Model testing the link between length of association (days) and the sex ratio within the brood, controlling for birth year and brood size. The tests indicate that length of association with parents (days) was influenced by brood sex ratio ($F_{5,325}$ = 2.75, P = 0.019), but when sexes were treated separately the effect of brood size remained only for male goslings ($F_{4,171}$ = 2.54, P = 0.042).

[47] Test of length of association (days) for males who were reared only with other male siblings compared to males reared only with female siblings (t test assuming equal variance t = 2.46, df = 114, P = 0.015). Males reared with other males associated with parents for shorter periods than males reared with females. The same comparison for females was not significant; test of average time with parents for females with only female siblings compared to only male siblings (t = -1.48, df = 97, NS). Length of association was not influenced by the independent variables (i) number of times a gosling was captured, (ii) location of initial capture, (iii) total geese in catch, (iv) goslings in the catch, and (v) number of times resighted in winter. None of the models for either sex approached the 0.05 level of significance in step-wise General Linear Model procedures (SAS 2001). Analysis included 192 females and 213 males.

[48] All three frequencies occurred more than would be expected by chance; binomial tests at least at the P < 0.04 level.

[49] An assessment of survival for goslings between the ages of three and 15 months was made using resighting records of 625 young Barnacle Geese. We asked if survival was a function of brood size, degree of contact with parents and length of association with parents. See Appendix 1, Table 4 for details of selected models.

[50] Length of association (days) with parents during first year and future reproduction. Males that eventually bred successfully stayed in the family with parents for longer than those that did not breed (t = -3.086, df = 57, P = 0.003; Figure 6.8). Length of offspring–parent associations did not differ for breeding and non-breeding females (t = -0.138, df = 59, P = 0.89). Unsuccessful males stayed with parents less than unsuccessful females (t = -1.867, df = 78, P = 0.066). Successful males and successful females stayed with parents for a similar amount of time (t = 1.117, df = 38, P = 0.27). Variation in age of first breeding (as assessed during winter) in relation to length of association (days) and brood size was tested with a General Linear Model procedure (SAS 2001) for one cohort (1986) of 61 females and 59 males. Length of association significantly influenced age of first breeding for males ($F_{1,58}$ = 5.22, P = 0.026), but not females. Brood size was included in the model, although this did not have a significant effect on age of first breeding. See Appendix 1, Table 5 for full details.

[51] Degree of contact with parents (proportion of observations with parents). The dependent variables, age of first breeding, lifetime reproduction from 1985 to 1995 and lifespan, were not influenced by the independent variable 'contact with parents' (controlling for birth year, brood size and gosling sex). None of the models approached the 0.05 level of significance in a step-wise General Linear Model procedure (SAS 2001). The analysis included 141 goslings that were observed with parents through the winter and into the spring period, with a minimum of five observations. Variation in this measure of association (proportion of observations with parents) did not vary among years, brood sizes or sexes.

CHAPTER 7
Nest parasitism, adoptions and kin

Sitting in a hide in the early morning, an observer watches hundreds of goose pairs nesting on the island. Some females are already sitting on eggs while their males stand vigilant. Other pairs are still in the process of establishing territories, comparing last year's location with alternative sites. Periodically, fights break out between adjacent pairs, but most of these are brief encounters. A lone female stands on the perimeter of the colony. She makes a short flight over the island and returns to the perimeter. Suddenly she moves quickly into the colony toward a particular nesting pair. She increases speed, throws herself on the nest and begins to push her way underneath the incubating female who attacks, walking on to the intruder's back and vigorously pecking and biting. Surprisingly, the nest owner's mate just watches the struggle take its course. The intruder female does not fight back. She lies still while adding one of her own eggs to the clutch. As soon as the egg is produced, the intruding female returns to the perimeter of the colony. However, because of the fighting, the newly laid egg has landed just outside the nest rim. After a brief examination the nest owner tucks the new egg under her bill and rolls it safely into the nest. She arranges the nest material and continues incubating. After observing a sequence of events like this we realise that a case of intraspecific nest parasitism has occurred.

Barnacle Geese practice social monogamy where pair members feed, migrate, defend the nest and associate with young in clearly identifiable social units (Chapters 4–6). This does not mean, however, that everyone in the brood is a true genetic offspring. In this chapter we describe behaviours that affect the occurrence of unrelated young in broods, including extra-pair fertilisations, intraspecific nest parasitism, and adoptions of newly hatched young. We investigate costs and benefits for individuals involved in these behaviours. As a next step, we explore whether goose societies are composed of larger social groupings – beyond the pair bond – by providing evidence for kin clustering in breeding colonies and long-lasting relationships among familiar individuals. We discuss functional aspects of these clusters and the potential benefits of having familiar neighbours in the colony.

Extra-pair fertilisations

Many studies on birds have shown that social monogamy is not the same as sexual fidelity (Griffith *et al.* 2002). Copulations between a socially monogamous female and one or more males outside the social pair bond may, depending, for example, on timing, give rise to extra-pair fertilisations. This in turn results in young that are genetically unrelated to the male that provides care for them. The frequency of extra-pair copulations can be determined by close field observations, but to reliably estimate the frequency of extra-pair fertilisations – that is, successful extra-pair copulations – one needs to use molecular genetic techniques. Extra-pair fertilisations occur in a large number of socially monogamous birds but the frequency varies among species, populations and individuals due to differences in life-history parameters and ecological factors. In geese, extra-pair copulations and extra-pair fertilisations have been found to occur at low or moderate frequencies in Lesser Snow Geese (Lank *et al.* 1989a), White-fronted Geese (Ely 1989), and in Black Brant (Welsh & Sedinger 1990). In our study of Barnacle Geese, we employed DNA fingerprinting and microsatellite techniques and found that extra-pair fertilisations were absent or very rare (Choudhury *et al.* 1993, Larsson *et al.* 1995, Anderholm *et al.* 2009a).

The absence or very low frequency of extra-pair fertilisations in Barnacle Geese might be due to the intense and vigorous mate guarding by the male partner before and during egg-laying. It has been debated whether mate-guarding behaviour during this period originally evolved in males to protect paternity or mainly to assure that his female partner effectively acquires nutrient stores to service egg-laying and incubation (Lamprecht 1989). Whatever the ultimate reason might be, mate-guarding behaviour apparently limited the possibilities for extra-pair males to copulate with paired females.

Although extra-pair fertilisations are rare in Barnacle Geese this does not mean that all eggs and goslings have necessarily been fertilised and produced by their social parents. On

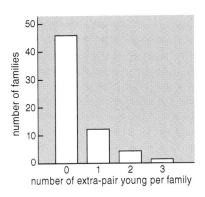

Figure 7.1. *Distribution of extra-pair young among 63 families in the main Baltic study population where goslings' parentage was assessed with DNA fingerprinting. From Larsson et al. (1995).*

the contrary, a DNA fingerprint analysis in the Baltic study population revealed that 17 (27%) of 63 analysed families had extra-pair young in them (Figure 7.1). In these cases, the approximately six-week-old goslings were not genetically related to either of their social parents (Larsson *et al.* 1995). There are three potential sources for these extra-pair young. Some may have originated from nest 'take-overs' at the beginning of the breeding episode. Others may result from intraspecific nest parasitism, an alternative reproductive strategy by which females lay eggs in the nests of others. Another likely source of these extra-pair young is from adoptions of unrelated goslings after hatching. Each possibility is discussed below.

Nest take-overs

Extra-pair young may result when pairs acquire a nest with a previous nest owner's eggs in it. This happens when pairs shift to a new nest location after laying an initial one or two eggs. The shift may occur either because a new site becomes available as the snow continues to melt or because pairs are forced out by aggressive neighbours. The original nest and eggs are then 'taken over' by prospecting pairs. In one year, for example, 10 out of 111 nests fit this classification in the Diabasøya colony in Svalbard. We have not quantified the survival of eggs that are taken over by new pairs, but it is potentially a source of extra-pair young. Nest take-overs have also occasionally been observed in the Baltic colonies.

Intraspecific nest parasitism

Intraspecific nest parasitism is a spectacular strategy that has evolved in over 200 bird species, including geese (Lyon & Eadie 2008). Parasitic females, or parasitic pairs, might increase their survival and reproductive success by (i) exploiting the often costly parental care provided by other unrelated individuals, (ii) spreading offspring among nests and thereby minimising the risk of loss of all progeny, and (iii) reducing competition between siblings (Andersson 1984, Petrie & Møller 1991). Behavioural strategies of hosts to minimise the probability of being parasitised, such as nest guarding, ejection of parasitic eggs and nest desertion, have also evolved in many species (Davies 2000). However, nest parasitism may not be costly for the hosts; for example, if the host pair and the parasitic female are closely related (Andersson 2001).

During our studies of the nesting colonies in the Baltic and in Svalbard we have identified several styles of parasitic egg-laying (Figure 7.2), including (i) laying of parasitic eggs in unguarded host nests, (ii) laying parasitic eggs in nests that are vigorously defended by host females, and (iii) leaving parasitic eggs just outside host nests that are subsequently retrieved by host females (Forslund & Larsson 1995).

In some cases, intraspecific nest parasitism can be detected by counts of eggs in nests. If more than one egg is laid on the same day, more than one female must be involved. Likewise, if a new egg appears in a nest many days after the nest owner has started incubation, another female must be responsible. The existence of extremely large clutches also indicates that more than one female laid eggs in the same nest, like the 'clutch' of 21 eggs that we once observed in a Gotland colony. We presume that clutches with eight eggs or more must have been

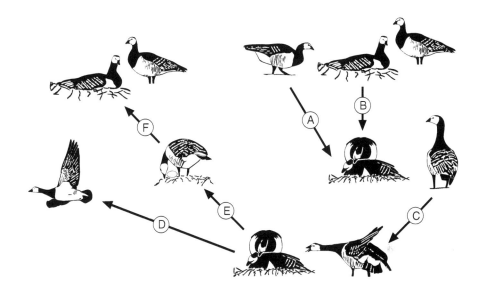

Figure 7.2. *Sequence of events during a parasitic egg-laying attempt. (A) The parasitic female rapidly approaches and tries to squeeze underneath the incubating female. (B) The host female responds by intensively attacking the parasitic female, pecking and biting at the head and neck. (C) The host male makes no or only a few attacks towards the parasitic female. (D) The parasitic female is passive and does not attack the hosts. After about 20 minutes the parasitic female lays an egg inside or just outside the nest and leaves. Sometimes a male joins her after the egg-laying attempt, but he does not interact with the nest owners. (E) The host female quickly rolls the parasitic egg (when laid outside the nest rim) into the nest cup. (F) After adjusting the eggs and nest material the host female continues incubation. Drawing by Kjell Larsson.*

produced by two or more females. Of course, smaller clutches may also contain parasitic eggs, but they can only be discovered by using DNA methods or egg protein fingerprinting, as explained below.

Eggs laid by parasitic females often end up on the nest rim, or even well outside the nest, and we have observed host females roll these parasitic eggs into their nests. One may wonder why they retrieve the parasitic egg just a few minutes after vigorously trying to get rid of the parasitic female. Parasitic females seem to exploit a stereotypic behaviour that has originally evolved to enable nesting females to retrieve their own displaced eggs (Lorenz & Tinbergen 1938). If not retrieved, such displaced eggs will obviously not hatch, and may increase the probability of nest failure by attracting egg predators to the nest (Lank *et al.* 1991). If parasitic females lay eggs in nests where the hosts have already completed their clutches and started incubation then the parasitic eggs will not develop in synchrony with the hosts' eggs and will therefore most likely not hatch.

The frequency of intraspecific nest parasitism in our main study colony in the Baltic was estimated by a non-destructive method based on egg albumen sampling and protein fingerprinting. After drilling a small hole in a newly laid egg a tiny albumen sample was taken with a syringe. The hole was sealed with glue and the egg developed and hatched normally (Andersson & Åhlund 2001, Anderholm *et al.* 2009a,b). Eggs laid by the same female have the same albumen protein fingerprint; that is, identical band patterns when the albumen proteins are separated on a polyacrylamide gel. The albumen fingerprints of eggs from genetically related individuals, for example mothers and daughters, resemble each other more than fingerprints from randomly selected individuals. By visiting nests during the egg-laying period, observing the exact position of the nests in the colony, noting the laying sequence of the eggs and performing protein fingerprinting we could estimate the number of parasitic eggs in clutches, analyse if parasitic eggs were laid before or after the host female had completed her laying, and estimate the fine-scale kin structure in the colony as well as the average degree of relatedness between parasitic females and host females.

Of 86 nests with complete information, 36% were parasitised, and 12% of all eggs were parasitic (Anderholm *et al.* 2009a). In that study year almost 80% of the parasitic eggs were laid after the host began incubation and the hatching success of these eggs was therefore limited. In an earlier study we found that about 55% of the directly observed parasitic egg-laying attempts occurred after the host had initiated incubation (Forslund & Larsson 1995). Although the frequency of intraspecific nest parasitism may vary among years and colonies and many parasitic eggs do not hatch because they are laid too late, it is clear from our studies that intraspecific nest parasitism is a frequently employed alternative breeding strategy in Barnacle Geese.

Age and experience of parasites and hosts

Some of the observed nest parasites and hosts had rings, which gave us information about their age and previous breeding experience. Parasitic behaviour might be limited to young females that are less able to establish nests of their own or to females without social partners (Andersson 1984, Lank *et al.* 1989b). However, in our small sample of 12 ringed parasitic females in the Baltic and Arctic all were at least four years old, and three were at least six years old. We know that eight females were paired and had nests of their own in years prior to laying parasitic eggs. Two of the parasitic females had nests of their own in subsequent years and four others had nests of their own later in the same year.

On some occasions we observed a parasitic male (defined as a male joining the parasitic female after the parasitic egg-laying attempt) standing some distance away from the host territory. These males were never observed to take an active part in the egg-depositing attempts. On other occasions the parasitic female was apparently alone, even though we watched for 10 minutes after she had left the host nest. Furthermore, our records confirmed that at least one of the ringed parasitic females was unpaired and at least two were paired during the breeding season.

The parasitic behaviour of Barnacle Geese seems to differ somewhat from the behaviour observed in Lesser Snow Geese. In that species, parasitic males sometimes help the parasitic female gain access to host nests by keeping the host male occupied (Lank *et al.* 1989a). Such behaviour has not been observed in our study populations. Furthermore, the possibility of switching between parasitic egg-laying and normal nesting seems to be less developed in Lesser Snow Geese than in Barnacle Geese (reviewed in Davies 2000).

Our observations of Barnacle Geese show that parasitic behaviour is not confined to one fixed type of bird. Some parasitic females have mates and nests of their own in the same or subsequent years and others do not. Nest parasitism has been described as an alternative strategy of lower success than normal nesting – that is, a 'best-of-a-bad-job' strategy – in

The Barnacle Goose is a long-lived bird forming life-long pair bonds. The oldest ringed bird in our ~dy populations was 27 years old. Photo by Kjell Larsson.

Barnacle Geese in spring. The rounded abdomen of the female, to the left, indicate that she has ~ained the necessary stores needed for egg-laying and incubation. Photo by Kjell Larsson.

▲ Svalbard in early June. In years with late snow melt, food availability is low throughout much of the incubation period. In such years only few eggs hatch. Photo by Jouke Prop.

▼ Barnacle Geese are specialised grazers and may take up to 300 bites per minute. Photo by Jouke Prop.

The Diabasøya colony in Svalbard. Barnacle Geese choose to nest on small barren islands, ccessible to Arctic Foxes. Photo by Jouke Prop.

Goose eggs are attractive food for Polar Bears. Few nests survive after several bears have visited a ony. Photo by Jouke Prop.

▲ White-tailed Eagle capturing a nesting adult Barnacle Goose. Increased predation on nesting adults and juveniles by eagles affects the size of Baltic colonies. Photo by Kjell Larsson.

▼ View over the main Baltic study colony on Gotland with several geese visible on the nest. The behaviour of ringed individuals was observed from hides. Photo by Kjell Larsson.

Intra-specific nest parasitism, that is, laying an egg in other´s nest, is a common alternative breeding strategy in Barnacle Geese. Here, a parasitic female has forced herself on a nest in an attempt to lay an egg. The parasitic female is vigorously attacked by the nest owner. Photo by Kjell Larsson.

A nest with newly hatched goslings in the Baltic. Photo by Kjell Larsson.

▲   Svalbard is an inhospitable place for Barnacle Geese when arriving in May. Slopes of seabird cliffs provide opportunities to recover from the long migration. Photo by Christiane Hübner.

▼ The Barnacle Goose is reliant on agricultural land for much of the year. Reducing economic damage by increasing numbers of geese may be a goal in management plans. Photo by Kjell Larsson.

▲   In the non-breeding season Barnacle Geese may aggregate in flocks of several thousands of individuals. Family members keep track of each other during seemingly chaotic events by using individually distinct calls. Photo by Kjell Larsson.

▲ Keen observers detect individual plumage differences, especially in the cheek patches. Rarely, white (leucistic) Barnacle Geese were observed. Photo by Inge Bakkeland.

▼ After all nests have hatched researchers visit offshore islands in Svalbard in inflatable kayaks (left). Even at this time of the year, the research camp may be covered by fresh snow. Photos by Jouke Prop.

years when breeding opportunities are reduced (Rohwer & Freeman 1989). Our finding that many parasitic eggs are laid too late relative to the host female's incubation period also indicates that nest parasitism in Barnacle Geese is a 'best-of-a-bad-job' strategy (Anderholm *et al.* 2009a). However, we cannot rule out the possibility that some high-quality females may try to increase their total reproductive output by combining nest parasitism with normal nesting in the same year.

Regarding the hosts, the available information from ringed birds indicates that the majority were at least four years old. Many of the hosts also bred before the year in which they were targets of parasitic egg-laying attempts. This indicates that the hosts were not necessarily targeted because of their inexperience. Future workers could assess whether parasitic geese actively choose to leave eggs with high-quality hosts that are more likely than others to succeed in hatching and rearing young (*sensu* Brown & Brown 1996).

Adoptions

Goslings younger than about six weeks that have been separated from their parents have little chance to escape attacks from gulls, eagles, foxes and other predators. Getting lost is easy in the first days, as young goslings encounter many situations where they stray away from parents. For lost goslings it is clearly beneficial to try to be adopted by other pairs, and any behaviour of young goslings that helps to find the protection of their parents or foster parents should be positively selected.

Adoption of foreign young is common in a variety of bird species (Eadie *et al.* 1988, Beauchamp 1997). In geese, adoptions are readily observed, for example, when two or more families come in close proximity and goslings mix in one group. When parents walk away with a different number of goslings than they had before the encounter, an adoption has occurred. Adoptions can also be detected by comparing a pair's original clutch size with their subsequent brood size, or by comparing brood sizes of marked pairs over consecutive days. Any increases over time must be due to adoptions. Adoptions may also be detected by genetic analyses of captured family groups. Adopted young, as well as young originating from intraspecific nest parasitism, will appear as genetically unrelated to both their social parents and siblings (Choudhury *et al.* 1993, Larsson *et al.* 1995).

Observations in both the Svalbard and Baltic populations showed that adoptions occurred regularly. On average, 17% of all goslings that hatched at the Diabasøya colony were adopted into another family and at least one adoption occurred in 21% of the broods (Table 7.1). Adoptions occurred especially during poor weather when several recently hatched families were waiting to leave the island to swim to the mainland tundra. On the mainland, adoptions still occurred, but were less frequent. While the majority of these adoptions (91%) occurred when goslings were less than a week old, and before individual recognition abilities had developed (Cowan 1973), the age of adopted goslings on the mainland ranged from 20 to 28 days of age.

Some pairs attending broods showed aggressive behaviours towards lone, contact-seeking young, which prevented the latter from coming close, but others seemed to readily accept additions to the family. From the adopting parents' perspective there may be both costs and benefits involved with adoption of foreign young (Avital *et al.* 1998, Kalmbach 2006).

	1980	1981
Total number of families	75	25
Total number of goslings	240	65
Average brood size	3.2	2.6
Adoptions: on nesting islands/on mainland tundra		
Number of families adopting goslings	10/1	6/1
Number of goslings adopted	14/1	16/2
Percentage of all families adopting goslings	15%	28%
Percentage of all goslings adopted	6%	28%

Table 7.1. *Exchange of goslings between broods at Diabasøya colony, Svalbard, during two summers of close observation. These figures only include successful adoptions. There were several cases where orphan goslings attempted to join families, but they were repeatedly chased away by the adults and subsequently taken by gulls. From Choudhury et al. (1993).*

In the first place there may be energetic costs, for example when a larger brood requires more costly vigilance behaviour from the parents, and when the new young eventually compete for food with family members. However, these costs are generally thought to be small and may be outweighed by advantages associated with the larger brood size. The presence of foreign young may reduce the probability of predation of their own offspring by dilution (Hamilton 1971), and the foreign young may later assist the host family to defend good grazing sites against other conspecifics (Choudhury *et al.* 1993). As social dominance is positively related to brood size (Black & Owen 1989b), adopters might benefit from accepting foreign young. Indeed, by first translocating newly hatched goslings between nests of marked pairs, thereby producing experimentally enlarged (addition of two young) and reduced (removal of two young) broods, and then monitoring the behaviour of experimental families, Loonen *et al.* (1999) found that enlarged families won more agonistic interactions at brood-rearing sites than reduced families. But how is that possible? Older goslings at the age of several months might assist their parents in encounters, but the small goslings in this study could hardly have the strength to actively help during an encounter with another family unit. Instead, the conclusion was that the presence of more goslings might affect the motivation of the parents and the opponents to engage in aggressive interactions.

Future success of extra-pair young and their social parents

To evaluate the possible costs and benefits of caring for extra-pair young – that is, young either originating from nest parasitism or from adoptions – long-term effects should also be considered. In the Baltic population we quantified the post-fledging fate of the 137 fledglings, which by DNA fingerprinting had been classified as: (i) within-pair young in families without extra-pair young, (ii) within-pair young in families with at least one extra-pair young, and (iii) extra-pair young (Larsson *et al.* 1995). Survival from fledging to arrival on the wintering grounds (post-fledging survival) was high in all categories and in both sexes. Actually, only 11 (8%) goslings were not observed in the subsequent winter or later and the resighting rate did not differ significantly between the three classes of young. We also found that the age of first breeding did not differ among gosling classes. Furthermore, survival and future reproductive success did not differ for parents with and without extra-pair young (Larsson *et al.* 1995). Therefore, any long-term costs or benefits of caring for extra-pair young after fledging seem to be small or are cancelled out.

Long-term social relationships with relatives

The complexity of social monogamy in geese is increased when one recognises that mating in the pair is not necessarily exclusive, or that not all the young in the family are true genetic offspring. In the previous sections we showed that unrelated individuals, whether parasitic females or unrelated young, exploited the parental care provided by socially monogamous pairs. Now we take the subject one step further and assess whether long-term social relationships exist between individuals beyond the pair bond, and whether such relationships affect the reproductive success of socially monogamous pairs. For example, if a parasitic female is genetically related to one of the hosts, or if neighbouring breeders in a colony are related, the complexity of goose societies is indeed increased.

In Chapter 6 we described family dynamics and how goslings in winter or spring decide, or are forced, to leave their parents. Before splitting up, some goslings spend 7 to 11 months together with their parents and brood mates and during this period they learn to recognise each other's visual and vocal cues. To examine the persistence of long-term social relationships between close kin after goslings leave their parents, we analysed the micro-geographic settling pattern of marked adult females and males that returned to breed in natal colonies in the Baltic (van der Jeugd *et al.* 2002).

We found that daughters that returned to breed in their natal colony nested closer to their parents than expected by a random distribution, but the nest location of sons conformed to a random pattern.[52] Sisters also nested closer to each other than expected, but only those that were born in the same year (Figure 7.3 and 7.4). Sisters did not nest closer to brothers born in the same or different years, and brothers born in the same or different years did not nest closer to each other than expected.[53]

Because the non-random settlement of sisters could be an effect of sisters settling close to their parents and thereby also close to each other, we examined the distance between nest sites of sisters that had settled on a different island to their parents. We found that these sisters also nested closer to each other than expected when born in the same year but, again, not when born in different years.[54] This settling pattern among sisters seems to indicate a genuine preference for nesting close to familiar kin, with whom they had close contact earlier in life. A possible mechanism explaining the clustering of sisters born in the same year could be that they settle as a team. However, this was not the case because the brood mate sisters that were found nesting close to each other did not always settle in the same year (van der Jeugd *et al.* 2002). Rather, we suggest that females actively sought out their parents and sisters when looking for a place to establish their first breeding territory and then settled close to them.

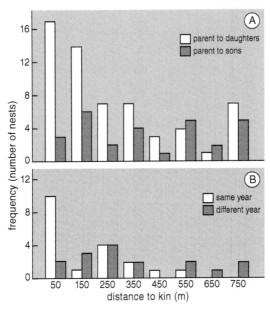

Figure 7.3. *Frequency distribution of the distance between nests of related Barnacle Geese. (A) Distance between nests of parents and nests of daughters and sons. (B) Distance between nests of sisters born in the same year and sisters born in different years. From van der Jeugd et al. (2002).*

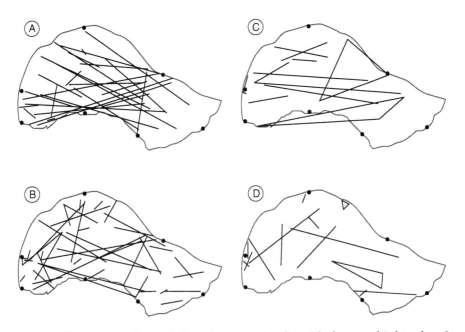

Figure 7.4. *Settling patterns of Barnacle Geese that returned to their 40ha large natal Baltic colony. Lines connect nest sites of related birds. (A) Distance between nest sites of parents and sons. (B) Distance between nest sites of parents and daughters. (C) Distance between nest sites for sisters born in different years. (D) Distance between nest sites of sisters that were born in the same year. Dots represent position of observation hides. From van der Jeugd et al. (2002).*

Future researchers may find it interesting to consider possible conflicts between males and females over how to behave towards neighbours and where to settle. As territories are established after pair formation, there must be some form of communication from the female to the male because the female apparently decides where a territory is to be established and the male takes the lead in establishing and defending this territory. The period of territorial establishment in goose colonies is full of threats, calls, and in some cases quite serious fights among neighbouring pairs until boundary lines are determined (Collias & Jahn 1959, Inglis 1977, Owen & Wells 1979). Perhaps these fights are less severe or absent in adjacent pairs that include kin.

Our knowledge about the mechanisms by which animals can recognise and discriminate between kin and non-kin in different social networks is growing (Kurvers *et al.* 2013). The data we have seem to imply that Barnacle Geese do not use a mechanism such as phenotype matching to recognise unfamiliar kin. Phenotypic matching involves learning the phenotype of familiar relatives, or of oneself, and thereby forming a template against which unfamiliar individuals can be compared (Komdeur & Hatchwell 1999). Rather, Barnacle Geese seem to recognise familiar individuals, which they probably know well from earlier in life (Choudhury & Black 1994, van der Jeugd *et al.* 2002).

During the period between family break-up in the goslings' first winter and when they breed for the first time at an age of two to five years, family members may have some contact. We periodically observed adult offspring together with their parents and adult siblings walking or foraging next to each other in winter flocks. In some of these cases the parents were also accompanied by young from a recent breeding season and the siblings were with their long-term mates. Such long-term associations during winter have also been observed in other geese (e.g. Warren *et al.* 1993, Fox *et al.* 1995, Frigerio *et al.* 2001). Whether these associations are mainly the result of philopatry to certain familiar winter grazing sites, or if the geese actively seek each other out to obtain some advantage, remains unexplored for Barnacle Geese.

There is no doubt, from our observations of individually marked birds, that females returning to breed in their natal colony nested closer, on average, to their parents and brood mate sisters than would be expected by chance. Further evidence for the existence of a fine-scale kin structure in breeding colonies was gained by a protein fingerprinting analysis of eggs in a large number of nests. Average relatedness was significantly higher between neighbouring females nesting less than 40 metres from each other than between females nesting further apart (Anderholm *et al.* 2009b).

Rates of intraspecific nest parasitism may be shaped by kin structure among colony members (Lyon & Eadie 2008). The occurrence of nest parasitism among kin is common in Goldeneyes *Bucephala clangula* (Andersson & Åhlund 2000) but avoided in Wood Ducks *Aix sponsa* (Sherman 2001). Low rates of nest parasitism among kin are expected if there is a disadvantage to the addition of foreign eggs (Sherman 2001). On the other hand, we might expect more nest parasitism among kin if it would result in larger broods, which subsequently engender higher dominance status and access to better foraging situations. The protein fingerprinting analysis in the main Baltic study colony showed that female hosts and parasites were on average no more closely related than randomly drawn nesting females in the population. However, although not significant, there was an indication that host–parasite relatedness was higher for timely parasites, which laid eggs within the host's laying

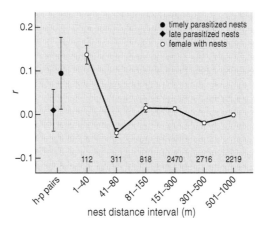

Figure 7.5. *Pairwise relatedness (average ± SE; number of pairs in class) in relation to nest distance for females with nests (n = 132) excluding parasites. 'h–p pairs' represents female host–parasite pairs, with timely (n = 14) and late (n = 25) parasitic laying. Relatedness between individuals was estimated by protein fingerprinting; that is, by bandsharing analysis of the egg albumen band patterns. From Anderholm* et al. *(2009b).*

sequence, than for late parasites, which laid eggs after the host female laid her last egg (Figure 7.5). To fully understand these patterns, additional behavioural observations of genetically related individuals are needed.

While the analyses which showed that some kin nest near to each other (van der Jeugd *et al.* 2002, Anderholm *et al.* 2009b) are intriguing and warrant further study, it was clear that many nesting neighbours were not genetically related to each other. It is likely that interactions between neighbouring geese, kin or non-kin, may be influenced by the degree of familiarity with each other (Kurvers *et al.* 2013). In Chapter 4 we showed how familiarity played a major role in the mate choice process; long-term partners were established from a pool of individuals that were reared in the same neighbourhood groups. Cooke *et al.* (1983), describing the structure of a Lesser Snow Goose colony, showed that each additional cohort of first-time nesters was grouped together on the edges of the larger colony. This clustering of same-aged birds also occurs in some Barnacle Goose colonies (I. M. Tombre unpubl. data.), leading to the possibility that some nest-neighbours may have been reared in the same neighbourhoods. In the next section we use data from the main study colony in Svalbard, and explore the ways in which neighbouring geese might cooperate during the breeding season.

Familiar neighbours

A large proportion of the geese in the Diabasøya colony were individually marked (> 65%), which allowed us to distinguish grouping patterns within and between years. Unfortunately, only a few kin relationships were known in this study area. Instead, we determined the degree of familiarity among group members based on a retrospective assessment. For this purpose, nesting locations of individual pairs collected over four years were analysed. Pairs were selected for analysis if they had bred in at least two of the years. A total of 53 pairs out of an average of 150 in the colony met this criterion. Using a cluster analysis technique we

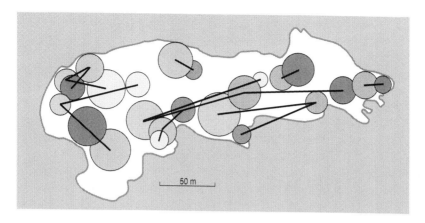

Figure 7.6. *Clusters of individual breeding pairs in the study colony on the island of Diabasøya, Svalbard. Circle size represents the average distance of nests to the cluster centre (each cluster is composed of 4–11 pairs). Locations of the same cluster in consecutive years are connected by lines. Note that not all clusters were present in each of the years and that clusters represent c. 30% of the breeding pairs. Later years are indicated by darker shading.*

aimed to distinguish groups of neighbouring pairs nesting near each other in consecutive years. The analysis identified eight groups, or clusters, of breeding pairs varying in size from 4 to 11 pairs (Figure 7.6). Only 2 of the 53 pairs were not assigned to a cluster.

Many of the pairs returned to the section in the colony where they had nested during the previous breeding attempt. Although geese rarely selected exactly the same location on the ground, 75% of the pairs (n = 53) returned to within 70m of the earlier nest site. This means that the clusters that were distinguished could have resulted from a fidelity to the site per se, rather than birds seeking the proximity of particular neighbours. From our results it is hard to distinguish between these two possibilities. We have two arguments, however, to support the idea that the clusters resulted from geese choosing to nest with familiar individuals. Firstly, although nest area loyalty was the rule, there were several exceptions. Two of the eight clusters moved considerably across the island between years (Figure 7.6), suggesting that a move of one pair incited the move of others. Secondly, we identified the composition of flocks on the brood-rearing and moulting sites by an intensive ring-reading schedule throughout summer (Prop *et al.* 1980). It appeared that associations found in the colony persisted for much of the summer, and that pairs of each group had their specific set of lakes to rear their broods and to moult.[55] Although we were not able to determine whether these groups were composed of kin, the observations provide evidence that geese may operate in groups with familiar individuals throughout the summer.

Finding familiar neighbours

There are several possible ways for neighbourhood groups to form in a colony. Familiar birds travelling together may settle as a unit. In this case, settling dates of cluster members would be similar, and variation among clusters would be large. The reverse was observed, however, as cluster members appeared to arrive and settle on different dates.[56] Thus, each of the clusters was created over a range of dates.

Cluster formation might be related to the relative ease of squeezing new territories in between those that are already being defended. Later arrivals face an ever-increasing problem of establishing a territory in the colony. Pairs that arrived late settled at closer proximity to other nests than early pairs, a consequence of filling up the colony island, and it took longer for them to establish a territory (Figure 7.7).[57] Assuming a causal relationship between the time needed to settle and available space in the colony (Figure 7.7C),[58] prospecting pairs might reduce trouble in establishing a suitable territory by seeking the proximity of relatives or otherwise familiar pairs. If true, this would imply that energetic considerations, i.e. the effort needed to establish a territory, might be the mechanism behind cluster formation in colonies. By reducing encounters with neighbours, goose pairs may also avoid the negative aspects of elevated levels of stress hormone (corticosterone), which are linked to other physiological processes and overall health (Scheiber *et al.* 2013).

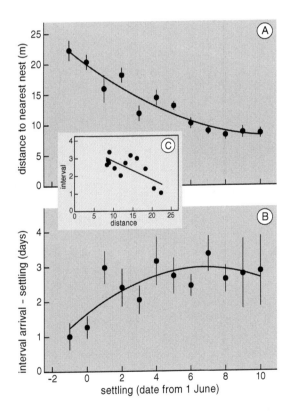

Figure 7.7. *Nearest neighbour distance between nests and nest establishment (settling date) for Barnacle Geese on Diabasøya, Svalbard. (A) Distance from nearest neighbour nests at the date of settling. Pairs arriving later in the colony settled at closer proximity to other nests. (B) Interval between arrival at colony and nest initiation (days to settling) across dates of settling. Late arriving pairs needed more time to settle. (C) Interval between arrival and settling and distance to the nearest neighbour across dates of settling. Longer intervals were associated with closer proximities to neighbours and vice versa. Average ± SE indicated. Based on 195 nests.*

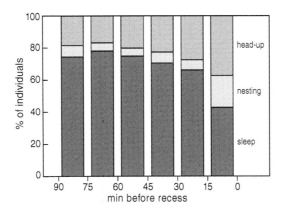

Figure 7.8. *Proportion of females that were vigilant ('head-up') in relation to time before initiating an incubation break ('recess'). The usual posture of incubating females was head-on-back ('sleeping'). 'Nesting' usually meant repositioning the down lining in the nest. Based on 720 records of 50 females.*

Potential benefits of neighbours

In addition to potential benefits like reduced fighting, individuals may obtain beneficial information about foraging opportunities from neighbours. Living in groups is often claimed to improve an individual's chances of finding appropriate foods (Pulliam & Caraco 1984).

In the study colonies on Svalbard, incubating females spent approximately one hour per day off the nest foraging on the tundra (Prop & de Vries 1993). The recession of snow, combined with the low density of food, caused a large variation in food availability through the season (Chapter 8). An individual goose attempting to find its daily ration of food was faced, therefore, with highly unpredictable foraging conditions. Foraging success could possibly be enhanced if a bird followed other individuals to preferred feeding areas. However, we did not find evidence to support the notion that individuals from the same cluster were together in foraging groups during incubation breaks more than could be expected by chance.[59] Rather than following each other during incubation breaks, they may have collected information from cluster associates while still on the nest. For example, from approximately one hour before an incubation break, females were increasingly looking around, in a head-up posture (Figure 7.8); during the rest of incubation, females spent most of the time sleeping, with their head on their back. This increase in vigilance may have had several functions; the most important, we speculate, was watching the other colony members, noting where they came from, how long they were off the nest and, on their return, determining their foraging success from the fullness of their oesophagus.

We hypothesise that geese may use this information to decide when and where to go foraging (Prop *et al.* 1984) and they may choose to nest near particular neighbours to capitalise on this information. During incubation in the Arctic, the Barnacle Goose diet consists of three main plant groups: grasses and sedges, herbs and horsetail, and mosses (Prop & de Vries 1993), each requiring a particular grazing technique to collect and ingest (Chapter 8). Females differed consistently in the food they selected, and we were therefore able to classify them according to items found in their diet. This showed that among clusters, large differences in diet choice occurred.[60] Although plant groups showed considerable overlap in distribution on the tundra, each had its main area of occurrence (Figure 2.3). Therefore, we suggest that a moss feeder would gain little information about the location of good moss areas from cluster members that specialise on herbs, and vice versa. If diet preferences were driven by a bird's ability to find and harvest only particular types of food then we might expect individuals with the same diet preferences to associate in space and time. Thus, while general 'foraging information' can be taken from any neighbour, specialised information about specific diets may be observed by nesting near particular individuals that form clusters. Information derived from neighbours would be particularly useful when it is complementary to the receiver's requirements, like a common preference of food plants. As compatibility between partners is an argument leading to long-term partnerships (Chapter 5), the same argument may hold for cluster members in a breeding colony.

Summary and conclusions

A DNA fingerprint analysis of 137 Barnacle Goose goslings revealed that 27% of the families contained extra-pair young at fledging. By definition, extra-pair young are unrelated to one or both parents. In Barnacle Geese, it is rare for extra-pair goslings to originate from copulations outside the social pair bond. Instead, the majority of these extra-pair young originated from intraspecific nest parasitism and adoptions. In a few cases, they may also have resulted from eggs that were incubated by pairs that 'took over' others' nest sites. Barnacle Goose females frequently employed the alternative reproductive strategy of nest parasitism; of 86 nests 36% were parasitised, and 12% of the eggs were parasitic. The majority of the parasitic eggs were laid after the host began incubation and the hatching success of these eggs was therefore limited. Parasitic behaviour was performed by females without mates or nests and by paired females with their own nests. Parasitic eggs often ended up just outside the nest, but host females rolled these eggs into their nests, possibly because unretrieved eggs may attract predators. After hatching, Barnacle Goose pairs also adopted foreign goslings; 17% of broods had at least one adopted gosling in them. Adoptions occurred, for example, during poor weather when pairs with recently hatched broods waited on the nesting island. Some pairs attending broods showed aggressive behaviours towards lone, contact-seeking young, which prevented them from coming close, but others seemed to readily accept additions to the family. As social dominance is positively related to brood size, adopters may benefit from accepting foreign young. Apparently, extra-pair young and their foster parents are not penalised in terms of survival and future reproduction.

In the assessment of long-term relationships beyond the pair bond, we found that females on average nested closer to their parents and brood mate sisters than expected by a random distribution. This indicates a genuine preference for nesting close to familiar kin, with whom they had contact earlier in life. Studying the interactions between familiar neighbours – kin or non-kin – is key to understanding the evolution of the spatial structure of colonies. For this purpose, we investigated foraging behaviour in relation to nest location in the colony. We speculate that nesting close to familiar pairs provides a foraging advantage for females by helping to decide when and where to feed.

Statistical analyses

[52] To test for non-random settling with respect to kin, randomisation tests were performed by comparing the actual average distance between nest sites of related birds to the distribution of average distances obtained by randomly allocating relatives to each other within the same sample 5,000 times (Adams & Anthony 1996). Females nested significantly closer to their parents than expected (Figure 7.3A; average distance 212m, n = 53, randomisation test: P < 0.0002), but nest sites of males conformed to a random pattern (Figure 7.3A; average distance 392m, n = 28, randomisation test: P > 0.2). Sisters also nested closer to each other than expected at random, but only when born in the same year (Figure 7.3B, same year: average distance 188m, n =19, randomisation test: P < 0.0002; different year: average distance 326m, n = 17, randomisation test: P > 0.3). From van der Jeugd *et al.* (2002).

[53] Sisters did not nest closer to brothers born in the same or in a different year, and brothers born in the same or different years did not nest closer to each other than expected at random (average distances 315, 362, 471 and 360m, n = 18, 13, 2 and 4, respectively, randomisation tests: all P > 0.3). From van der Jeugd *et al.* (2002).

[54] When settling on a different island, sisters born in the same year nested closer to each other than expected if they had settled at random (average distance: 135m, n = 10, randomisation test: P < 0.0002), but there was no tendency for aggregating among sisters born in different years (average distance 347m, n =11, randomisation test: P > 0.5). From van der Jeugd *et al.* (2002).

[55] Contingency table consisting of pairs by cluster (eight clusters) and brood-rearing and moulting areas (four areas): χ^2 = 38.5, df = 21, P = 0.01. Data from Prop unpubl. data.

[56] ANOVA of dates of settling, averaged for each pair, by cluster: $F_{7,43}$ = 0.91, NS.

[57] ANOVA of nearest neighbour proximities across dates of settling ($F_{2,192}$ = 27.1, P = 0.001). See Figure 7.7A. ANOVA of time needed to settle across dates of settling ($F_{2,163}$ = 5.47, P = 0.005). See Figure 7.7B.

[58] Regression of time to settle (interval between arrival date and settling date) in relation to distance to nearest neighbour (r = -0.71, n = 12, P = 0.01). See Figure 7.7C.

[59] Tabulating the number of foraging trips on the tundra with another individual from the same cluster or not (females only): χ^2 = 6.12, df = 7, NS.

[60] MANOVA of diet composition by cluster: $F_{21,111}$ = 1.93, P < 0.05.

CHAPTER 8

Food and feeding

Geese spend most of the daylight of each season collecting food, and in some situations foraging continues during moonlit nights. This perceived gluttony gives the impression that they live in a world of plenty, but nothing is further from the truth. For much of the year geese are recovering from periods of depleted body stores, or they are preparing for energy draining activities (also see Chapter 11). Geese have to process large amounts of food to be able to extract the energy and nutrients they need. Moreover, weather conditions, predators or other sources of disturbance regularly affect foraging activities. As a consequence, only the most efficient foragers will achieve an enhanced potential for reproduction and survival.

This chapter is about the tactics and strategies geese use to find their food. We will describe foraging behaviour in relation to food abundance and quality with the aim of better understanding how many birds certain habitats can support (i.e. carrying capacity). In particular, we explore whether individuals vary in their preferences and abilities for acquiring food and whether feeding sites are monopolised by a select group of birds.

Barnacle Geese are incredibly selective in choosing where to feed. They directed most foraging activity on vegetation patches with the highest density of grass shoots, as seen from numbers of droppings left behind on the foraging grounds (Figure 8.1). Spending most time in the densest patches presumably yields a higher intake of food (see below). That droppings were also found in areas with low grass densities indicates that the birds sampled these less profitable sites but spent little time there before moving on. This relationship between goose grazing and food supply raises some important questions. The linear relationship between densities of droppings and shoots indicates a consistent ratio of grazing pressure to the amount of food (1.75 droppings per 1,000 shoots), but does this ratio represent the maximal grazing pressure possible, or could additional geese have visited the sites? Was exploitation

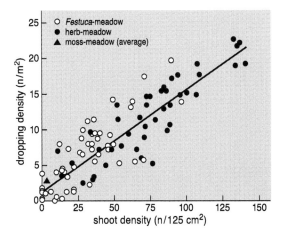

Figure 8.1. *Relationship between dropping density and grass shoot density in various meadow types in Helgeland, Norway. The regression line is for all three vegetation types combined, showing that geese spent most foraging time on plots with highest shoot densities.[61] Dropping density values accumulated over the season within 4m² plots. Shoots were sampled in subplots. Data restricted to zones of suitable grazing habitats.*

based on dominance relationships or due to experience gained over the years? These questions will be addressed in this chapter, starting with the introduction of the 'functional response', which provides a basis for understanding foraging decisions.

Functional response

The relationship between food intake rate and the amount of available food is referred to as the 'functional response' (Holling 1959). Typically, the rate at which food is obtained (referred to as intake rate) increases with higher food density, and levels off to reach an upper limit at highest densities. Levelling off is caused by the time needed to handle food. In predators, handling time is often related to swallowing the prey *after capture*, and in avian herbivores it mainly represents the time to process the food in the bill and oesophagus *after swallowing* (Crawley 1983).

To explore foraging behaviour of Barnacle Geese in relation to food density, we marked nine plots of 5×5m on a *Festuca rubra* meadow in the spring staging area in Norway. Plots were located to represent the variation in food abundance in the area. In herbivore studies, total amount of green biomass is often used as a measure of food availability (Drent & Prins 1987). However, for selective grazers, like geese, lumping all plant parts into one category would lose important information. A more precise set of measurements included density of shoots, number of leaves per shoot, and the size of leaves before and after goose grazing occurred. Foraging performance was quantified by recording peck rates and the proportion of time spent feeding. Bite sizes were determined by comparing before-and-after sward measurements (Chapter 3).

Figure 8.2 shows how geese adjusted their foraging tactics according to the state of the vegetation. Peck rates and the number of leaves taken per peck increased with shoot

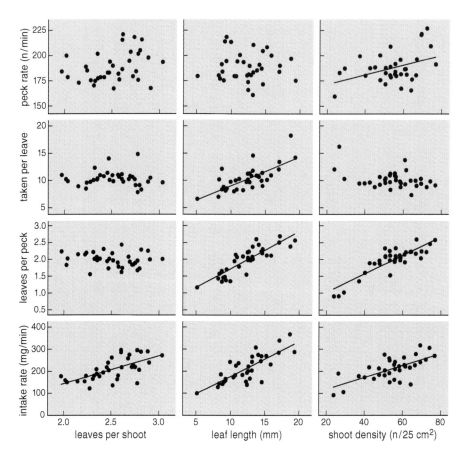

Figure 8.2. *Foraging performance of Barnacle Geese feeding on the grass* Festuca rubra *in Helgeland, Norway. Food availability identified by leaves per shoot (left), leaf length (middle) and shoot density (right). Intake rate of food (bottom) derived from peck rate (top), size of leaves taken (second from top) and number of leaves per peck (second from bottom). Each component of the sward (columns) had a significant effect on food intake rate (rows).[62] Partial plots presented, in which effects of other variables have been accounted for.*

densities, and the number and size of leaves taken increased with leaf size. Overall, intake rate was positively related to all three characteristics of the sward; in order of importance: length of a leaf, number of leaves per shoot, and density of shoots.[62]

However, there was no evidence that food intake rate levelled off at highest food densities, which was expected on theoretical grounds (Holling 1959). The reason for the persistent linear trends may be that data were collected in early spring, when handling time was apparently not limiting because food densities were low. Van der Wal *et al.* (1998) showed that intake rates of captive Barnacle Geese kept in pens on a saltmarsh in the Netherlands levelled off, or even decreased, at much higher food densities than those encountered in our study area.

It is expected that the functional response varies among plant species (Durant *et al.* 2003). Barnacle Geese feeding on mosses, for example, may be able to collect large bites, but due to long handling times intake rates of moss eaters are relatively low (Prop & de Vries 1993). Subtle differences in handling time may also exist among grass species due to variation in the angle and size of leaves. The functional response is further expected to vary among individuals, due to variation in bill shape (Durant *et al.* 2003) or the bird's age (Lang & Black 2001). The act of grazing by geese involves a complicated interplay of movements of the bill, tongue and pharynx, which are performed at incredibly high speeds. Details of foraging mechanics are easily overlooked when observing geese in the wild, as indicated from an analysis of feeding behaviour captured by high-speed camera (Box 8.1). These observations allow an appreciation of the agility and flexibility of goose foraging techniques.

BOX 8.1 MECHANISM OF GOOSE GRAZING

In this box we describe some of the subtleties of goose grazing. Each peck into the food layer comprises a set of intricate bill movements and head angles before a food item is grasped and eaten. The neck is extremely flexible, enabling geese to move their heads in almost any position, apparently to effectively crop a variety of food types. For example, grass leaves are grazed by keeping the head slightly tilted such that the angle between the bill and ground (the pronation) is approximately 70°. Large leaves of, for example, Sea Plantain *Plantago maritima* are approached in a perpendicular fashion (90°), whereas stalks of Variegated Horsetail *Equisetum variegatum*, which are often lying flat on the ground, are taken when the head is almost parallel to the ground (20°).

While it is possible to judge the angle of the head and neck while watching geese graze, head and bill movements are often too fast for an observer to distinguish details. Yet they are revealing, and understanding the fine details of peck movements makes it easier to appreciate real time grazing. Addy de Jongh with John Videler used high-speed film cameras (100 frames per second) from two different positions to record Barnacle Geese feeding on a grass turf, thus achieving a three-dimensional picture of head and bill movements (de Jongh 1983). On average, a single peck took less than half a second (x-axis, Figure 8.3A). The downward movement (the approach) took approximately a quarter of this time; another quarter was required to grasp the leaf while the head was at its lowest position in the sward. During the remaining time the head was lifted and brought into position for the next peck. The bill was opened wide on the way down and then closed at the bottom of the approach (Figure 8.3B). On grasping a leaf, a goose did not immediately tear it off. Instead, the bill was opened slightly and closed again. This subtle movement served perhaps to reposition the leaf in the bill, to have it pointing backwards, or to squash the leaf to facilitate breaking. A combined movement of the head and bill accomplished the actual cutting of the leaf; while the bill was closed, the head moved up and at the same time the tip of the bill was moved rapidly towards the body (Figure 8.3C). The subsequent fine movements of the bill are synchronised with movements of the tongue that transport food particles from the tip of the bill into the direction of the oesophagus. When Barnacle Geese ingest large amounts of food items, the 'handling time' increases, and consequently the peck rate drops. As the bill continues to open and close at a high frequency, this mode of feeding resembles filter feeding in dabbling ducks. Multiple items are collected just at the opening of the oesophagus to form a neat package before swallowing (van der Leeuw *et al.* 2003). When geese take a break between feeding bouts, often to scan the surroundings, the bolus of food is pushed downwards into the oesophagus. After a bout of one minute, the goose in Figure 8.3 would have a bolus of 150 leaves.

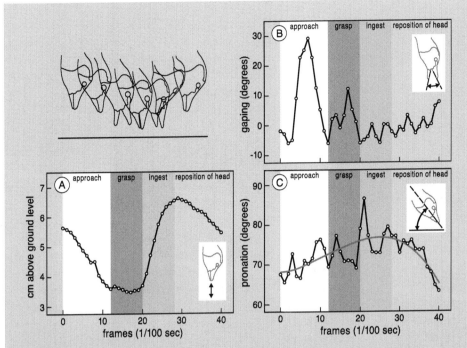

Figure 8.3. *Mechanism of rapid grazing actions in Barnacle Geese. A single peck movement on a 5cm tall grass sward is described from analysis of a high-speed camera sequence (100 frames per second). Indicated are the (i) approach, (ii) grasp, and (iii) ingestion of leaves plus repositioning of head. The leaf taken broke at frame number 20. (A) Vertical position of the head above ground surface. (B) The angle between the lower and upper bill (gaping). (C) Angle between the bill and the ground surface (pronation). Drawings represent digitised versions of the original film frames. From de Jongh (1983).*

Depletion and phenology of food plants

Geese are choosy feeders, often focusing on only a few species at a time when foraging in diverse plant communities. The distribution of food plants is usually very heterogeneous, occurring in zones or discrete patches exhibiting large variation in densities, which forces individuals with particular preferences to visit multiple locations to get what they need. Barnacle Geese select for the most profitable parts of a plant, taking the younger and larger leaves when feeding on grasses (Figure 8.4), even when pecking at extremely high rates. Due to the gregarious nature of geese, many individuals will visit feeding areas that have been previously grazed (Sutherland & Allport 1994, Clausen & Percival 1998, Rowcliffe *et al.* 2001). Prop & Loonen's (1989) study of Brent Geese foraging on saltmarshes showed that depletion of the main food plants may occur at surprisingly high rates. The first three individuals visiting patches of Sea Plantain removed most of the leaves available, and very little was left at the patch for individuals at the rear of the flock. Analysis of film records of Brent Goose flocks combined with assessments of food available in micro-plots showed that 50% of the available plantain was cropped by only 12% of the birds in the flock, and 27%

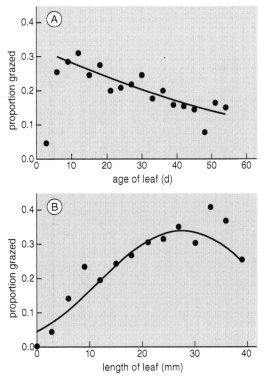

Figure 8.4. *Intensity of grazing (proportion of leaves grazed) of* Festuca *in relation to (A) age and (B) size of leaves in the spring staging area in Helgeland, Norway.*[63]

of all individuals did not find any plantain. What happens if geese come back a couple of days after this severe grazing? It is in these situations of heavy depletion of food plants that geese become increasingly dependent on regrowth of grazed vegetation and emergence of new plants (see below).

In spring and summer in particular, plants exhibit rapid changes in growth and ageing, which cause a dynamic pattern in availability of goose food. The spring flush of plant growth means a sudden burst of food becoming available. However, new plants may be available for only a short time before becoming unattractive to herbivores. Buds of Polar Willow *Salix polaris*, for example, are an attractive food item in early summer (Prop *et al.* 1984). However, the high nutritional value of the plants is much reduced when buds develop into leaves containing high concentrations of phenolics and condensed tannins (Dormann 2003), a process that takes just a few days. As a consequence, sites that are profitable at a certain moment may become unattractive quite soon after.

Another example of geese having to track the coming and going of their food involves Purple Saxifrage *Saxifraga oppositifolia* flowers, which brighten the tundra during the early breeding season in Svalbard. The flowers are a favourite food for Barnacle Geese on incubation breaks (Prop & de Vries 1993) because of the nectar that is produced during sunny days (Hocking 1968). It is not a surprise that during cloudy and misty conditions – which are common in the Arctic – the flowers are avoided by geese.

Foraging tracks and intake rate

Building on the theme of the previous section, we ask how geese manage to find their daily ration. Is it by finding and exploiting profitable feeding areas, or by fighting for and monopolising preferred sites? Observations were made in the spring staging area in Norway. The study site was located on the Lånan 'home island' (Chapter 2), for which a detailed vegetation map was available. Geese visited the island each morning (03:00–10:00 hours); some pairs spent all their time in the observation area, others a variable portion of the morning. The position of goose pairs and singletons were plotted at 10-minute intervals throughout the staging period. Males often chased adjacent pairs while their females foraged, and some pairs seemed to monopolise small patches of vegetation. Based on the proportion of aggressive encounters that were won, the 'dominance status' of marked pairs (one or both partners ringed) in the study area was quantified.

Three parameters were derived from these records. First, the sequence of positions of marked pairs showed the route that was travelled through the observation area. These tracks enabled us to calculate the net distance moved per unit time (the 'rate of walking'). Secondly, for each marked pair, the proportion of time spent per vegetation type was calculated. This was achieved by comparing positions with the vegetation map. In this analysis we focused on vegetation that provided the highest intake rate, i.e. 'rich patches'.

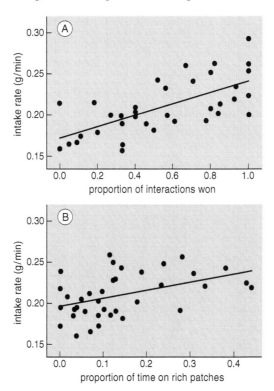

Figure 8.5. *Intake rate by spring staging Barnacle Geese in relation to (A) proportion of aggressive encounters won, and (B) proportion of time on rich vegetation patches.[64] Partial plots presented after removing effects of other variables.*

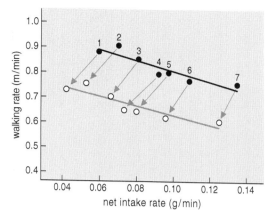

Figure 8.6. *Walking rate of foraging Barnacle Geese in relation to net food intake rate (intake rate multiplied by digestibility of food) during spring staging in Norway. Seasonal averages for seven vegetation types (closed symbols). Walking rate varied among vegetation types; geese moved slowest in vegetation providing the most and best food.[65] Open symbols represent situation when grazing pressure increased by 30 minutes per m² causing a decline in net intake rate and walking rate. Arrows connect data points of corresponding vegetation types: 1 = muddy, 2 = mosses, 3 = mosses/Festuca, 4 = Festuca, 5 = Puccinellia, 6 = Poa, 7 = rich patches Festuca + Poa.*

These patches were distinguished by a lush growth of grasses. Thirdly, because unmarked individuals were also plotted, it was possible to calculate the grazing pressure per square metre that accumulated through time, giving the prior grazing pressure for the position of each pair at any time. Intake rates for each pair were estimated from the individuals' dropping rates (mass and intervals) and the digestibility of food. To explore individual variation in intake rates we regressed them against the three factors described above and the birds' dominance status.

Individuals that ingested food at the highest rate were those who won most aggressive encounters and spent most time on rich patches (Figure 8.5). Apparently, it paid to be aggressive, although the analysis also showed that aggression and patch use were not always linked; some individuals spent much time on rich patches, and achieved a high intake rate, without being aggressive at all.

It may be expected that prior grazing pressure would have an additional negative effect on intake rate, due to depletion of intensively grazed areas. This was not the case, however, as intake rate was independent of prior grazing pressure.[64] Apparently the decline in food availability could be compensated for by reducing the speed with which individuals travelled through depleted areas. Geese adjusted walking rates to local foraging conditions in two different ways (Figure 8.6). They slowed their walking rates down, or they made more turns, while on the best sites where intake rates were highest. Moreover, when food availability dropped in the course of the season, geese responded by walking slower, which was apparent in each of the vegetation types. All individuals followed this pattern, but there was considerable variation, as some birds moved consistently faster than oth This indicates that individuals used different foraging strategies – i.e. some geese w and others were apparently more deliberate foragers – which is an intriguing asr birds' personalities.

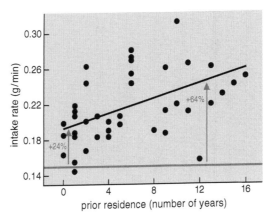

Figure 8.7. *Relationship between intake rate and prior residence of individually marked geese that visited the Norwegian spring staging islands for 0–17 previous years. Intake rates improved with additional number of years in attendance at the site.[66] Indicated is the percentage increase relative to maintenance food intake (horizontal grey line) for novices (0–1 year residence) and veterans (10+ years). Age of birds was not associated with intake rate.[66]*

Individual foraging success

Results presented in the previous section suggest that successful foragers combine effective competitive behaviour with the ability to find the most profitable feeding spots. Who are these successful birds? Extensive ring reading in the study area from the mid-1970s onwards (Gullestad *et al.* 1984) provided useful information about the history of visits of individual pairs to the spring staging islands. The number of years of prior residence in the study area provided a key to describing foraging success (Figure 8.7). The influence is particularly striking when compared with the average intake rate needed to maintain body mass, 0.15g per minute (Prop *et al.* 1998). Novices in the area achieved an average intake rate of just 24% above this minimum value, and some individual pairs did extremely poorly considering that they were preparing for long-distance migration. In contrast, the intake rate of veterans was no less than 64% above minimum requirements.[66] Although prior residence and age of birds were closely correlated, there was no indication of a significant association between intake rate and age.[67] This suggests that 'local knowledge' adds more to successful foraging than the general experience gained with age, and that individuals benefit from returning to sites they have used before.

To further investigate the relationship between 'local knowledge' and foraging performance we analysed how individually marked pairs used the foraging area. Each day at sunrise the geese arrived in small flocks from the roost. However, this did not mean they operated as flocks. When looking in detail at the distribution of specific pairs within a flock, there were measurable differences in the way they moved over the island; each pair had a preferred part of the feeding area (Figure 8.8). On subsequent days, pairs usually followed the track taken on the previous day. For example, both AEU plus partner and the pair coded as CD/CM focused foraging efforts on the higher parts of the island, though each in a separate corner. In contrast, YJL plus partner foraged on the lower parts of the island, systematically exploring the rich patches (compare with the vegetation map, left bottom panel of Figure

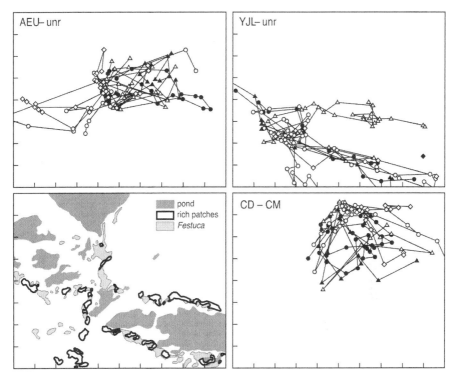

Figure 8.8. *Tracks of three pairs in observation area during spring staging in Helgeland, Norway. Selected are six consecutive days. Note consistent use of the area. Based on records at 10-minute intervals of individually marked pairs. Map (bottom left) indicates some of the key features of the plot.*

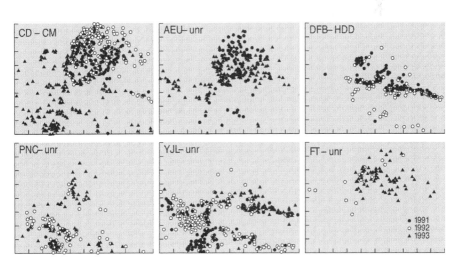

Figure 8.9. *Site selection of six individually marked pairs in three consecutive years during spring staging in Helgeland, Norway. Note consistent use of the focal area among years. Based on records at 10-minute intervals of individually marked pairs.*

149

8.8). Even more striking was that comparing the records over years revealed a very consistent pattern in the pairs' choice of sites. Pairs generally returned to the same part of the feeding area each year (Figure 8.9). However, they exhibited a subtle but important fine-scale shift in distribution from one year to the next. Apparently, as a result of the annual shifts pairs spent an increasing proportion of time on rich food patches.[68] Additional analyses gave the same result; i.e. the proportion of time on rich patches was positively related to the number of years of prior residence in the study area.[69] Furthermore, in statistical models the birds' age was inferior to 'prior residence' in explaining variance. We suggest, therefore, that pairs were able to improve their foraging performance across days and years by the sequential discovery

BOX 8.2 SOCIAL DOMINANCE AND FORAGING ADVANTAGE

The gregarious nature of wild geese is best seen on the wintering grounds where flock sizes number in the thousands. Flocks come daily to forage on refuge pastures, landing first in the middle and then radiating out to the edges. Social groups, composed of singles, pairs and family members, jostle for position while walking and foraging at different rates (Carbone *et al.* 2003). The familiar concept of a pecking order, where subordinate individuals are submissive to the more aggressive, is also realised in goose flocks (Boyd 1953, Raveling 1970, Lamprecht 1986a). By recording the location of social groups within foraging flocks, we found that the amount of food that was available to individuals was correlated with their dominance status. We found that due to depletion from repeated use of the pastures there was, on average, 91% more biomass on the perimeters than the centres of fields and that larger social groups tended to use these areas by positioning themselves on the edges of flocks (Table 8.1). Families were found more often on the leading edges, whereas non-family adults (usually in pairs) and single geese frequented the centre and other edges of flocks. Birds on leading edges of flocks probably obtained food of better quality since they were grazing plants that had not already been picked through by others. From examining plant fragments in goose droppings, we found that the food taken by edge-birds was more digestible and nutritious than that taken by birds in the centre of flocks (Black *et al.* 1992).

Table 8.1. *Percentage distribution of social classes among flock positions in Scottish pastures. Family size was positively correlated with time spent in the best positions (leading edge) and negatively correlated with time spent in the worst positions.*[71] *Updated from Black & Owen (1989b).*

	Leading edge	Centre	Other edge	n
Single goslings	52	19	29	132
Non-family adults	39	33	28	6,710
Single parent + goslings	44	35	21	75
Pair + 1 gosling	75	10	15	669
Pair + 2 goslings	79	5	16	456
Pair + 3 goslings	83	2	15	265
Pair + 4 goslings	85	< 1	14	142

and continued use of rich foraging spots. These observations provide a functional basis to goose 'traditions', i.e. site fidelity maintained over the years.

As experienced individuals spent more time in the best foraging sites, they would presumably suffer an increasing pressure from other geese competing for the same locations. However, the proportion of interactions that pairs won was not related to prior residence or age.[70] Apparently, pairs were well able to cope with increasing social pressure, and we speculate that gaining information on the location of better sites was linked to the bird's social abilities to defend them.

In early winter, when most goslings are still in family groups, the situation is quite different as dominance relationships are primarily determined by the size of social units. Box 8.2 further develops the concept that social dominance leads to the acquisition of prime foraging positions using an example from large flocks on the wintering area.

Limitation on numbers – carrying capacity

Herbivores are selective in what they take and where they feed, thus ensuring they ingest sufficient food of adequate quality. Consequently, they crop only a proportion of all food available (Figure 1.1), leaving less attractive plant parts untouched. In mechanistic terms, grazers employ a *lower threshold of acceptance* below which consumption is not rewarding (Drent & Prins 1987). The amount of food produced in an area and the lower threshold are among the factors that determine the herbivore density an area can support. This density is synonymous for 'carrying capacity' of the area (Ebbinge *et al.* 1975, Goss-Custard & Sutherland 1997). A first indication that an area is filled to capacity comes from an examination of annual counts; capacity is reached when numbers in the focal area remain constant over time even when numbers for the entire population change (Newton 1998). Individuals arriving at saturated locations or those experiencing depleted habitats may shift to alternative locations (the 'buffer' concept, Gill *et al.* 2001). Shifting to buffer areas may take place on any spatial scale (Gill *et al.* 2001) and movements are thought to be permanent – contrary to our findings explained below.

Barnacle Geese may use foraging areas to capacity in several periods of the year. For example, within the traditional parts of the spring staging area in Norway goose numbers remained remarkably constant from the 1970s onwards while the total Svalbard population continued to increase (Prop *et al.* 1998). This suggests that these areas had filled to capacity. High densities of geese also occurred at places with highest production of food (Figure 8.1), providing additional evidence that carrying capacity was reached.

Trends in distribution in the Svalbard summer range provide another example of habitat saturation. Comparing goose numbers with habitat maps showed that during the brood-rearing and moulting period the prime moss meadow habitat around tundra lakes supported a maximum of 10 geese per hectare (Prop *et al.* 1984). In the 1970s not all lakes had attracted large numbers of geese, but subsequently the growth of the local population was accommodated in lake systems that had apparently not yet filled to capacity (Figure 8.10). During the early 2000s, numbers continued to increase, this time by spreading to lakes that initially were thought to be unsuitable as brood-rearing habitat due to the large distance from the breeding colonies (Drent & Prop 2008). In particular, a large number of

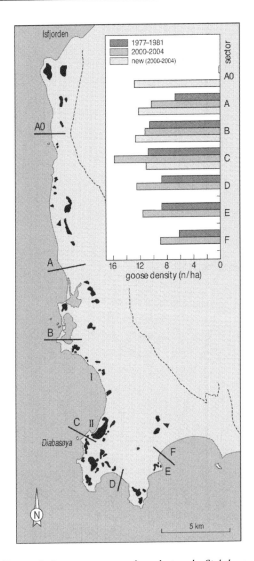

Figure 8.10. *Density of Barnacle Geese on moss meadows during the flightless period, Nordenskiöldkysten, Svalbard. Densities are maximal numbers observed in each sector divided by the area of moss meadows. A comparison is made between two periods (1977–1981 and 2000–2004). Increasing numbers were accommodated by filling low-density areas (for example sector A), and by expansion to new lakes. Based on Drent et al. (1998), Prop & Drent (2003), Prop & de Fouw (2004).*

lakes in the northern part of the study area were colonised in recent years. Whereas numbers in the study area increased almost fivefold since the 1970s, maximal goose densities on moss meadows during brood rearing remained quite stable.

However, more detailed examination of the birds' behaviour and interaction with food plants is required to understand if limitation of goose numbers is related to food conditions. The following sections describe situations during spring migration and brood rearing, exploring the birds' lower threshold of acceptance and the importance of buffer habitats, two concepts that provide a more comprehensive understanding of limitation on numbers.

Lower thresholds

We collected measurements on springtime food exploitation in Helgeland in a series of *Festuca rubra* meadows. At this time of year, grazing pressure on the plants was high (on average 48% of the grass production was removed), which led to a severe reduction in leaf lengths (Figure 8.11A). During the second half of the staging period, when depletion was most severe, goose grazing and plant production were in balance, as evidenced from a constant rather than an increasing food availability (Figure 8.11B). Geese accomplished this by visiting the plots each day and taking leaves that had grown to an exploitable size. This strategy enables maximal cropping rates because leaves are grazed before they become inedible due to ageing, a situation also found in other goose grazing systems (Prop 1991, Rowcliffe *et al.* 1995). How goose flocks achieve this fine-tuning of grazing is discussed in a following section.

Apparently, the intensity of grazing within plots was adjusted to the density of grass shoots (Figure 8.12A); high-density plots received the highest grazing pressure and low-density plots the least. Consequently, the initial relationship between leaf size and shoot density with smaller leaves at the highest densities (due to density-dependent growth in the *Festuca* sward) was maintained during the goose staging period.[72] In the first part of the staging period the highest intake rates were achieved in plots with the highest shoot densities (Figure 8.12B). However, after the continued visitation to the best plots, intake rates during

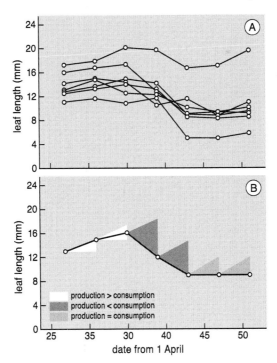

Figure 8.11. *Size of Festuca leaves as a measure of food availability during spring staging in Helgeland, Norway. (A) Seasonal trends for separate sample plots. (B) Averaged patterns across all plots, indicating relative importance of plant production to goose grazing. During the second part of the staging period food production and goose grazing were in balance.*

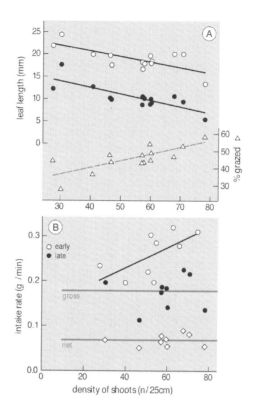

Figure 8.12. *Food availability and intake rate of food in sample plots during early and late periods of the spring staging period in Helgeland, Norway. (A) The negative relationship between leaf size and shoot density of Festuca which existed in the early period was maintained by a stronger grazing pressure (percentage of shoots grazed) in high-density plots.[72] (B) The positive relationship between intake rate and shoot density in the early period levelled off during the late period.[73] The constant net intake rate (intake rate corrected for digestibility) during late staging suggests a lower threshold of acceptance.*

the final part of the staging period dropped to a flat average value of 0.17g per minute, which we see as a lower threshold of acceptance (gross intake line, Figure 8.12B). When accounting for variation in food quality among plots by calculating the net intake rate of food, variation in the lower threshold is even further reduced (the coefficient of variation drops from 21% to 18%, Figure 8.12B). That net intake rates were consistent over the gradient of available shoot densities may mean that geese employ a lower threshold of acceptance incorporating both (gross) intake rate and food quality.

Buffer habitats

Adjusting the intensity of grazing to the amount of food available (Figure 8.11) suggests that goose flocks distribute over the feeding grounds in response to fluctuations in depletion and regrowth from one day to the next. This response by flocks must be accomplished by appropriate decisions by individual pairs within them. A key moment of a goose day is in

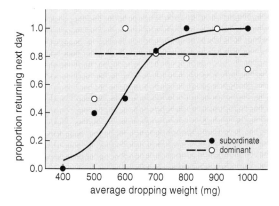

Figure 8.13. *Proportion of 115 pairs returning to the central 'home island' foraging area the next day in relation to foraging performance (intake rate approximated by average dropping weights). A relationship was apparent in subordinate pairs; whereas most dominant pairs returned the next day, only the subordinates with highest intake rates had a high probability of returning. Data from Helgeland, Norway.*

the early morning when pairs decide where to go for a new foraging session. The decision is thought to be based on the success that was achieved in the previous day (Ydenberg *et al.* 1983). In the Norwegian study, geese could choose from a variety of foraging locations, including the central, preferred home island, or – as an alternative – one of many peripheral islands with a less profitable plant cover.

We tested whether the previous day's foraging performance influenced the next day's decision by examining intake rates of marked individuals that visited the central island. The birds' probability of returning to the home island the next day was high after a successful day of foraging (using dropping weight as a proxy for intake rate; Figure 8.13), and after a poor day the return rate was low. Instead of returning to the home island, the geese fed on the peripheral islands, a relationship that was particularly strong in subordinate pairs. Home island return rates for dominant pairs were invariably high and independent of foraging success in the previous day.[74] These observations suggest that numbers on the feeding grounds were adjusted to foraging conditions by subordinate pairs returning or not. This corroborates the findings by Raveling (1969b) and Ebbinge (1992) who showed that site fidelity of geese on the wintering grounds was greater for dominant pairs than for subordinate individuals. Our results indicate that decisions take place on a day-to-day basis, resulting in a close tuning of grazing pressure to available food. Peripheral 'outer' islands served as a buffer habitat during periods when plant growth did not compensate for grazing pressure on preferred islands.

In summer, during the brood-rearing and moulting period, goose flocks used a similar complex of preferred and buffer habitats. Besides moss meadows with grasses and sedges, geese visited the fjellmark, a mesic tundra at a larger distance from the water (Figure 2.4). Though the fjellmark covered large areas and contained rich patches, food densities were on average lower than in moss meadows. Moreover, foraging on the fjellmark meant an increased risk of predation by Arctic Foxes. Initially, goose flocks grazed on moss meadows only, but when grazing pressure (as goose-day per hectare per day) exceeded one, an increasing proportion of time was spent at a larger distance from the water's edge (Figure

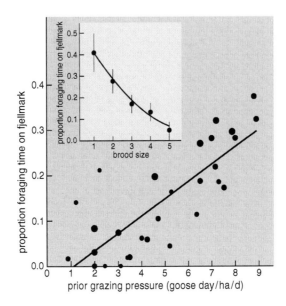

Figure 8.14. *Relationship between proportion of time spent on the fjellmark and cumulative grazing pressure on moss meadows on lake shores of Svalbard brood-rearing areas. Size of circles is proportionate to length of observation. Heavy grazing of moss meadows was compensated for by spending more time on the fjellmark 'buffer' zone.[75] The insert shows that time spent on the 'buffer' was related to brood size, with families with small broods spending most time on the fjellmark.[76] Averages ± SE indicated.*

8.14). The smallest families were the first to exhibit a shift to the alternative habitat, resulting in a negative relationship between brood size and proportion of time spent on the fjellmark (Figure 8.14). These observations are in line with the picture developed in Box 8.2 of larger families outcompeting smaller families by occupying the most profitable (and in this case safest) places to feed. Every 3.8 days (n = 20) a new grazing cycle started when flocks moved to a neighbouring lake, thus giving vegetation at the earlier lake time to recover. Overall, geese responded to brief periods of inadequate food availability by shifting to the buffer habitat.

Summary and conclusions

This chapter describes tactics and strategies that geese use to find their food. Foraging success depends on the structure of the vegetation; intake rates by Barnacle Geese were related to separate components of the sward, like length of leaves and density of shoots. Based on the relationships established it was possible to predict intake rates from vegetation measurements. Each peck was composed of rapid movements by the head, bill and tongue, which enabled the geese to ingest food at high rates even when feeding on short vegetation. Dense goose flocks may rapidly deplete food, which leads to large asymmetries in diet composition among individuals; the most profitable items were obtained by a minority of the flock members. The availability of food followed a dynamic pattern, as plants were usually suitable as food for only a restricted period. In the Arctic summer, the period between the appearance of food items and those food items developing beyond a palatable stage was often less than two weeks. Foraging success was closely associated with social behaviour (those that won most encounters achieved highest intake rates), and with movements over the foraging area (locating rich food patches). The effects of depletion by earlier flock members were compensated for by walking slower. Individuals improved foraging success with each year of returning to the same area. Novices performed poorly, even when they were old birds, whereas veterans (returning for 10 or more years) achieved highest foraging success. These observations suggested that local experience was a key to improvements in foraging success. Individual pairs tended to return each day to a particular segment of the foraging area. Over years, individual pairs returned to the same segments, though they moved gradually towards more profitable parts of the habitat. These observations of tenacity to foraging locations and improving foraging success provide a functional basis to site fidelity.

Goose numbers in an area may reflect the 'carrying capacity' of the area, which results from food production and a lower threshold of acceptance by the geese. Geese may occupy habitats and exert high intensities of grazing such that food consumption keeps pace with production of food plants. Buffer habitats were important to accommodate part of a local population when food production suddenly dropped. Subordinate individuals were most likely to make these temporary shifts to buffer areas.

Statistical analyses

[61] ANCOVA of dropping density (y) by shoot density (x, covariate) and vegetation type (factor). The effect of shoot density was independent of vegetation type (testing for differences in slopes, $F_{2,162} = 1.43$, NS), and the additive effect of vegetation type in a model with shoot density as a covariate was not significant ($F_{2,164} = 2.48$, NS): $y = 1.93 + 0.144x$, $n = 167$, $r^2 = 0.72$, $P < 0.0005$. From Prop et $al.$ (1998).

[62] Multiple regression of intake rate with number of leaves per shoot, leaf length and shoot density: $F_{3,31} = 27.2$, $P < 0.00005$, $r^2 = 0.72$. T-values for each parameter separately: 3.89 ($P = 0.0005$), 4.57 ($P < 0.0005$) and 3.05 ($P < 0.005$), respectively (Prop & Oosterbeek unpubl. data).

[63] Logistic regression of proportion of $Festuca$ leaves grazed with leaf length ($\chi^2 = 53.3$, df = 1, $P < 0.0005$), length^2 ($\chi^2 = 25.2$, df = 1, $P < 0.0005$) and leaf age ($\chi^2 = 20.1$, df = 1, $P < 0.0005$) as covariates. Based on repeated measures of leaves nested within sample plots. Data were from 10 to 20 May (Prop & Oosterbeek unpubl. data).

[64] Multiple regression of intake rate of individual pairs with proportion of interactions won, proportion of time on rich patches, and walking rate (and its squared value): $F_{4,32} = 14.0$, $P < 0.0005$, $r^2 = 0.63$. T-values for each parameter separately: 4.89 ($P < 0.0005$), 2.42 ($P < 0.05$) and 2.38 ($P < 0.01$), respectively. Prior grazing pressure was dropped from the model ($t = 0.05$, NS).

[65] GLM of walking rate by vegetation type ($F_{6,3507} = 4.45$, $P < 0.0005$) and individual pair ($F_{60,3507} = 3.04$, $P < 0.0005$) as factors. The interaction between vegetation type and individual pair was not significant ($F_{231,3276} = 1.05$, NS), underlining the consistent differences between pairs. The relationship between walking rate (y) and grazing pressure (x) (analysed by individual pair) was: $y = 0.963 - 0.0055x$, $F_{1,67} = 13.18$, $P = 0.0005$.

[66] Regression of intake rate and prior residence of individually marked geese: $y = 0.194 + 0.004x$, $F_{1,37} = 13.85$, $r^2 = 0.27$, $P = 0.001$. Comparing the intake rate of novices (prior residence 0–1 year, n = 10) and veterans (10+ years, n = 9): $F_{1,17} = 21.67$, $P < 0.0005$.

[67] Correlation coefficient between prior residence (years) and age of birds: $r = 0.685$, n = 39, $P < 0.0005$. Correlation between bird age and food intake rate: $r^2 = 0.049$, n = 41, NS.

[68] Comparing proportion of time on rich food patches between successive years by individual pairs: slope of the regression $0.046 \pm$ SE 0.030 per year, repeated measures ANOVA: $F_{1,31} = 4.66$, $P < 0.05$.

[69] Multiple regression of proportion of time on rich food patches with prior residence (years) and age: $y = 0.074 + 0.007 \times$ prior residence, $F_{1,80} = 7.49$, $P = 0.008$; age was not included in model, $t_{79} = -1.20$, $P = 0.23$. Analysis after randomly selecting one record per individual.

[70] Proportion of interactions won was not related to prior residence ($t_{34} = 1.12$, $P = 0.27$) or age ($t_{34} = 0.39$, NS).

[71] Contingency table of frequency of aggressive encounters for social groups sizes (seven) in different flock positions (three) at WWT Caerlaverock ($\chi^2 = 837$, df = 12, $P < 0.001$). Spearman Rank Correlation between social group size and proportional use of best flock position (leading edge; $r_s = 1.0$, n = 5, $P < 0.01$) and worst flock position (centre; $r_s = -1.0$, n = 5, $P < 0.01$). See Table 8.1. From Black & Owen (1989b).

[72] Leaf length of *Festuca* was negatively related to the density of shoots; both before arrival of geese (apparently a density-dependent effect in vegetation) and after severe grazing (slope of regression $F_{1,25} = 34.0$, $P < 0.0005$). Slopes of regressions were similar for both periods ($F_{1,24} = 0.34$, NS), resulting from a relatively heavier grazing pressure at high shoot densities. See Figure 8.12.

[73] ANOVA: During the early staging period intake rate of food was related to the density of shoots ($F_{1,7} = 3.94$, $P = 0.08$), but the relationship disappeared during the last 10 days of staging ($F_{1,7} = 0.08$, NS). See Figure 8.12.

[74] Logistic regression of the proportion of pairs returning the next day in relation to average dropping weight as an index of intake rate. All pairs ($\chi^2 = 10.9$, df = 1, n = 115, $P = 0.001$); dominant pairs (top-ranking 50% of all pairs; $\chi^2 = 0.0$, df = 1, n = 59, NS); subordinate pairs (lowest-ranking 50% of all pairs; $\chi^2 = 9.7$, df = 1, n = 50, $P = 0.002$). The difference in slope between dominant and subordinate pairs was significant ($\chi^2 = 7.6$, df = 1, n = 109, $P = 0.006$). The seasons' last observations were excluded from analyses. See Figure 8.13.

[75] Logistic regression of the proportion of time spent on the fjellmark in relation to prior grazing pressure of moss meadows near lake shoreline ($\chi^2 = 77.5$, df = 1, $P < 0.0005$), and date ($\chi^2 = 23.8$, df = 1, $P < 0.0005$). See Figure 8.14. Based on observations of 30 flocks (T. M. van Spanje, S. Dirksen & J. Prop unpubl. data).

[76] Logistic regression of the proportion of time spent on the fjellmark in relation to brood size ($\chi^2 = 16.9$, df = 1, $P < 0.0005$). Based on observations of 15 families during the second half of the brood-rearing period (S. Dirksen & J. Prop unpubl. data). See Figure 8.14.

CHAPTER 9

Survival and reproduction

In previous chapters we developed the concept that goose societies are composed of individuals with different competitive abilities that acquire unequal amounts of essential resources. In this chapter we will show that, although many try, surprisingly few actually succeed in transferring those resources into surviving goslings. A goose laying a viable five-egg clutch each year of a long life could potentially produce 100 offspring, yet in reality the most productive individual in our study only had 21 goslings surviving through autumn, and the majority of geese never produce any recruits at all.

Individuals of many species face a trade-off between breeding successfully and surviving to breed again in the future because breeding can be expensive and dangerous, resulting in 'costs of reproduction'. Like other animals, geese must deal with this dilemma between breeding and survival. Prudent decisions regarding the timing and relative effort in reproduction may be critically important (Drent & Daan 1980, Stearns 1992). Furthermore, the stochastic nature of weather conditions in the Arctic may heighten the costs involved. This chapter includes an investigation into the likelihood of survival and reproduction for Barnacle Geese with different life histories. We ask why so few individuals are successful in raising any offspring, and which scheduling strategy over a lifetime yields most success. Are more offspring produced by those who start breeding early in their lives, those who spread out their breeding effort, or those who postpone breeding till later in life? We also address how the increasing number of birds in our population has affected the relationship between breeding and survival.

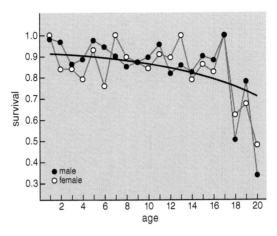

Figure 9.1. *Survival probabilities for the 1972 Barnacle Goose cohort by age. Symbols represent results of a capture–mark–recapture analysis to estimate annual variation in survival.[77] Solid line based on weighted averages of the models, showing a decrease in survival with age. Based on 46 males and 44 females captured as yearlings in 1973 in the Dunøyane, Svalbard breeding area.*

Survival and timing of mortality

The annual survival rate for geese in the Svalbard population was on average 0.90, beyond the first summer. The average lifespan was 9.5 years and the oldest goose lived for 27 years. Figure 9.1 shows that the likelihood of survival decreased with age, from over 0.90 in the first five years, to less than 0.85 at age 12, and dropping to below 0.70 for birds older than 20.[77] Survival for males was slightly better than for females. In the following sections we propose reasons for this pattern of senescence and the disparity between the sexes.

Figure 9.2 plots separate survival probabilities from an analysis including breeding status.[78] Females accompanied by young in the current winter had a lower survival in the subsequent year than those that did not have offspring. In other words, females who were successful breeders in the previous summer period, as indicated by their continued association with offspring in the winter season, were less likely to survive the ensuing 12 months (i.e. return again to the wintering area) after another season in the Arctic. Interestingly, male survival was independent of their breeding status. We suggest that the apparent effect of breeding status on annual survival is due to the cumulative costs associated with producing and caring for the brood. Representing survival from one winter to the next, as is shown in Figure 9.2, may actually include reproductive costs that are accumulated during consecutive summers and migratory periods.

The survival rate of adults from the Baltic population was estimated at an even higher rate of 0.95 (based on over 100,000 winter resightings of 2,336 birds between 1984 and 1997; Larsson *et al.* 1998), perhaps reflecting a less strenuous lifestyle that comes with their shorter migration. Survival for other protected populations with long-distance migrations include 0.86 for Dark-bellied Brent Goose, 0.87 for Light-bellied Brent Goose, and 0.88 for Russian Barnacle Geese (citations in Madsen *et al.* 1999). These values are considerably higher than values from hunted goose populations. For example, average adult survival was estimated at 0.84 for Black Brant (Ward *et al.* 1997), 0.83 for Eurasian White-fronted Geese *Anser*

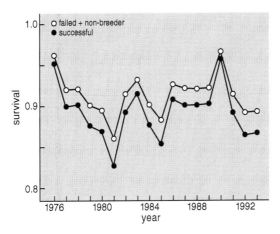

Figure 9.2. *Survival probabilities of Barnacle Geese (n = 3,429 females, 3,428 males) and breeding status, assessed in the winters of 1976–1993. Successful breeding was determined by counting goslings (> four months of age) that associated with marked birds after arrival on the Scottish wintering grounds. The survival model included year and breeding status as parameters.*

albifrons albifrons (Ebbinge 1991), 0.83 for Greater Snow and Pink-footed Geese (Gauthier *et al.* 2001, Madsen *et al.* 2002), 0.80–0.85 for Lesser Snow Geese (Francis *et al.* 1992), 0.75 for Greater White-fronted Geese *Anser albifrons frontalis* (Schmutz & Ely 1999), and 0.71 for Canada Geese (Hestbeck & Malecki 1989). Survival values for the non-migratory Barnacle Goose population in the Netherlands dropped from an initial value of 0.98 to less than 0.75 after a management hunt in summer was introduced (van der Jeugd 2013).

In spite of extensive shoreline searches in the winter area, we found very few dead bodies, usually less than a dozen in a year. For example, in 1993, when the Svalbard population numbered 14,350 birds, the estimated 12% mortality for that year should have resulted in 1,750 carcasses. In fact, most mortality of juvenile and adult geese is realised during the

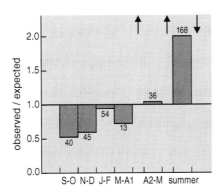

Figure 9.3. *Dates when Barnacle Geese were seen for the last time as an index of mortality during the annual cycle (n = 356 adult birds). Index expressed as the observed/expected number of losses based on the probability of sighting geese in each period, including September–October (S–O), November–December (N–D), January–February (J–F), March–first half of April (M–A1), second half of April–May (A2–M) and June–August (summer). Arrows indicate periods of northern and southern migrations between Scotland, Norway and Svalbard. From Owen & Black (1991a).*

migratory and summer periods and relatively few deaths occur during the non-migratory period on the wintering grounds. Figure 9.3 shows the time of year that 356 marked birds died based on their last resighting dates. Many more birds than expected (values above 1) disappeared in the summer period, including both migrations, compared to the autumn and winter months.

Mortality as 'cost of reproduction'

If a 'cost of reproduction' is measurable in a population, individuals that succeed in breeding may not be the same individuals that live for a long time, and vice versa. Hence, only certain combinations of survival and reproduction are possible in practice, and it may be these that set an upper limit to lifetime reproductive success (Partridge 1989). Therefore, to better understand what makes a successful goose and the limitations individuals face, in this section we describe the nature of the costs involved in reproduction.

There are apparent costs related to the stresses of gaining and losing body mass, and migrating first with an excess (in spring) and later with limited body stores (in autumn). This argument is a common theme in this chapter. For example, of 115 moulting females which had brood patches when we caught them in summer in Svalbard, indicating that they had recently incubated eggs, 105 (91%) were resighted in autumn back in Scotland, whereas all but one of the 47 females (98%) without brood patches survived the journey (Owen & Black 1989a). This suggests that individuals that did not breed, or that had abandoned the clutch in an early stage of incubation, survived better than those that had gone through a long period of incubation. This agrees with more detailed assessments of survival of different classes of birds that we observed in the breeding colony. Female survival from one breeding season to the next decreased with increasing reproductive investment: non-breeders survived at a rate of 0.95, failed breeders at 0.86, and successful breeders at only 0.82 (Prop *et al.* 2004). Male survival, on the other hand, did not change with breeding status, indicating that the roles performed in reproduction were substantially different between the sexes, influencing survival probability of females and less so that of males.

In the following sections we consider potential influences on reduced female survival associated with reproduction: (i) the effort of withstanding prolonged periods of fasting during the 25-day incubation and (ii) the costs of parental care during brood rearing. A third source of variation in survival, due to costs related to the scheduling of reproduction and moult, is discussed in Chapter 11.

Costs of incubation In an effort to quantify costs involved in reproduction Tombre & Erikstad (1996) conducted an experiment designed to alter the energy that female Barnacle Geese expend during incubation. They added or subtracted five days of incubation by swapping entire clutches that were due to hatch at different times. Female geese will readily accept foreign egg additions. This clever manipulation effectively created a disparity in reproductive investment. Some females had to remain on their eggs for five extra days, while others hatched five days early. Inserting eggs that were due to hatch at the normal time also created a set of control nests. Results from the experiment help us understand how females cope with the cost of incubation. Females that were forced to invest more in incubation lost more body mass than the other females. However, this extra burden was

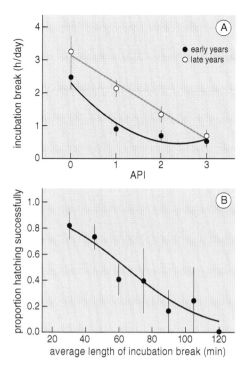

Figure 9.4. *(A) Relationship between the length of incubation breaks by female Barnacle Geese (hours per day) and the abdominal profile index (API) as a measure of body condition. The negative slope means that Barnacle Geese spent more time off the nest to forage when body condition declined. The slope of the regression was steeper in late snowmelt seasons, which means that foraging effort increased at a higher rate in late years when body stores were depleted. (B) The proportion of females successfully hatching a clutch was inversely related to the average length of an incubation break; the earliness of the snowmelt did not have an additional effect. Averages ± SE are indicated. After Prop et al. (1984).*

not enough to reduce the likelihood of survival to the next breeding season or influence the probability of initiating a nest in the next season. This experiment indicates that females were able to cope with extra demands during incubation without any immediate 'costs of reproduction'. This suggests that Barnacle Geese follow a strategy unlike the one described for Lesser Snow Geese, where body reserves may be depleted during incubation to such an extent that some individuals starved to death (Ankney & MacInnes 1978). In Barnacle Geese, as body stores became depleted during the incubation period females compensated by increasing the time off the nest to feed (Tombre *et al.* 2012). This response was most evident when comparing the situation in early and late seasons. Food availability was low in late years with a delayed snowmelt (Prop & de Vries 1993), and due to the lower food availability in late years females were found to spend more time off the nest than in early years (Figure 9.4A).[79] Since the probability of hatching was negatively correlated to the length of incubation breaks (Figure 9.4B), a female's response to diminishing body reserves reduced her chances of breeding successfully. Apparently, when conditions are poor female Barnacle Geese avoid the risk of mortality and sacrifice current breeding attempts in favour of future reproduction.

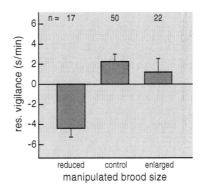

Figure 9.5. *Average residual vigilance of parent Barnacle Geese for three manipulated brood categories (reduced, control and enlarged) in an Arctic brood-rearing area. Residual vigilance differed between the experimental treatments. Parents of experimentally reduced families were significantly less vigilant than parents of experimentally enlarged or control broods. There was no significant difference between the control and the enlarged group. Average brood size was 1.4, 2.8 and 3.2 for reduced, control and enlarged broods, respectively. From Loonen et al. (1999).*

COSTS OF PARENTAL CARE Loonen *et al.* (1999) designed an experiment to test the effect of brood size on parental effort. They manipulated brood size by removing or inserting two newly hatched goslings. Depending on gosling age, parent geese will accept foreign goslings (Chapter 7). Parents that were given extra goslings spent more time standing guard in head-up vigilant posture than those that had a reduced brood (Figure 9.5).[80] However, this extra effort was not at all costly to the parents with regard to body condition at the end of the brood-rearing period. Female parents with the largest broods actually recovered more

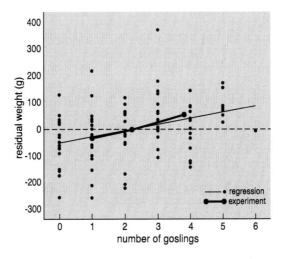

Figure 9.6. *Body condition of parent female Barnacle Geese during the moult in Svalbard, expressed as the residual body mass after correction for size (first principal component PC1), in relation to brood size. Small dots represent females caring for unmanipulated broods. Body mass of adult females was positively related to brood size (indicated by the solid thin line). The solid thick line connects average residual mass for three experimental categories where brood size was manipulated at hatch (reduced, control and enlarged broods). From Loonen et al. (1999).*

mass during the moult than those with smaller broods (Figure 9.6),[81] which suggests better prospects for survival. Indeed, the probability of survival and nest initiation in the next year was the same across experimental treatments, indicating that the extra burden of enlarged broods was not deleterious. However, in another breeding area in Svalbard female Barnacle Geese caring for 'natural' large broods did suffer a decline in survival during that summer compared to those with smaller broods; survival rate of females with one or two goslings was 0.94 compared to 0.85 for females with three or more goslings (Prop *et al.* 2004). We suspect that this disparity is brought on by differences in predation pressure by Arctic Foxes in these areas; an idea that warrants further study.

Annual reproductive success

Barnacle Geese are physically capable of reproduction at the age of two years, though most young birds did not attempt to breed until their third or even fourth year (Prop *et al.* 1980); the average age of establishing a first nest ranged from 2.6 to 3.6 years (Figure 9.7). Once reaching a mature age, the majority (80%) attempted to nest each summer, but only a small proportion succeeded and returned to Scotland with goslings due to many losses experienced along the way (Prop & de Vries 1993). For example, in the Svalbard study, only 50% of the nests hatched one or more of the eggs (successful nests assessed over a five-year period) and 29% of the pairs that hatched one or more goslings (successful in summer, n = 174) were not observed with offspring in the wintering area. For the population as a whole, only 17% (SE 0.02, n = 22 years) of pairs returned to Scotland with a brood (average 1.86 goslings, SE

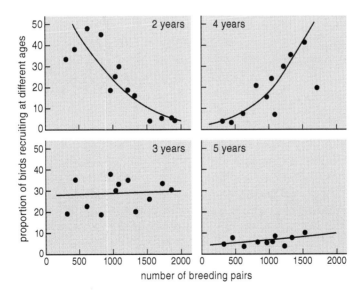

Figure 9.7. *Age at which female Barnacle Geese initiated their breeding careers in the main study colony (recruitment) on Gotland, Sweden. In females, the average age of first reproduction increased as number of breeding pairs in the colony increased. At high densities they postponed their first breeding attempt until the age of three or four. From van der Jeugd & Larsson (1999).*

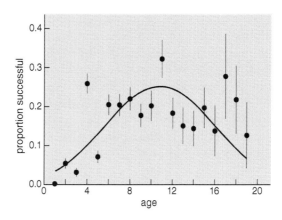

Figure 9.8. *Proportion of female Barnacle Geese bringing at least one gosling to the Scottish wintering area in relation to age (y = −3.84 + 0.51x − 0.024x² on a logit scale). The plot indicates an increasing reproductive success until age 10–11, after which fewer offspring are produced. Data for males are not presented but the relationships were similar. Averages ± SE are indicated. Modified after Black & Owen (1995).*

0.02, n = 2,832 marked females). This means that the vast majority of pairs that we observed foraging in the large wintering flocks will have flown to the Arctic, established a breeding territory and laid a clutch of eggs, but returned to Scotland without goslings.

Whether geese returned to the wintering grounds with goslings varied with age, increasing during the initial years, peaking at age 10–11 and declining in the older age classes (Figure 9.8).[82] The improvement in the early years was attributable to an enhanced performance at each step of the reproductive cycle gained from experience. The decline, apparent in both males and females, most likely resulted from the inevitable wear and tear of breeding and competition, which affected a combination of fitness parameters (Black & Owen 1995).

To investigate whether annual reproductive success (as determined in winter) changed over time, we considered three sets of geese that hatched on the same Svalbard coastline but during different years (1976, 1980 and 1986 cohorts). Snowmelt and weather conditions were not substantially different in these years, but competition for food during brood rearing is likely to have been more severe in the later years as the number of geese using the area tripled over this period, increasing from 1,000 to almost 3,000 individuals (Drent *et al.* 1998). With increasing goose densities, reproductive success, controlling for differences due to the birds' ages, decreased for the local population (Figure 9.9).[83] Although each cohort followed the same general pattern between years, a closer look at the trends in time revealed noticeable differences. The decline was strongest for the latest, 1986 cohort, whereas females of the earliest 1976 cohort maintained a more or less constant level of success. This difference is remarkable because each cohort apparently responded to an increasing density in a different way. An explanation for this may be that the 1976 birds maintained their success by becoming established breeders before the area became densely populated during the 1980s. Once established, and due to their larger body size (Chapter 10), the 1976 cohort was apparently better able to compete for the limited resources. The later cohorts may have been less able to become established breeders when numbers rose and competition for nests and food intensified.

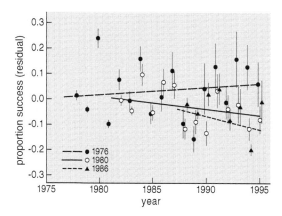

Figure 9.9. *Proportion of female Barnacle Geese bringing at least one gosling to Scotland for three different Svalbard cohorts (1976, 1980, 1986 Nordenskiöldkysten brood-rearing area). The analysis shows that trends in reproductive success (corrected for age) differed among cohorts. Averages ± SE are indicated.*

Lifetime reproduction

Lifetime reproductive success, the sum of offspring produced in a lifetime, is a useful variable for describing selection in a population (Clutton-Brock 1988, Newton 1989). There are three components to this composite variable: size of broods, the proportion of years breeding successfully, and lifespan. Individuals that start breeding at the earliest age, produce the largest broods and succeed in each year of a long lifetime achieve maximum lifetime reproduction. However, if there are costs associated to any of the breeding parameters

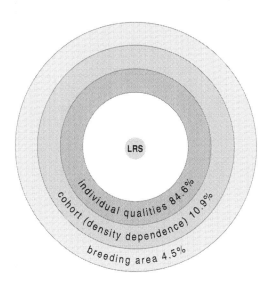

Figure 9.10. *Factors influencing female lifetime reproductive success in the Svalbard Barnacle Goose population, listed hierarchically: breeding area, cohort, and individual qualities. Percentages of the variance associated with each factor (derived from an ANOVA) are indicated. Components of lifetime reproductive success are given in Figure 9.11.*

the picture becomes more complicated. For example, the cost of obtaining a territory and producing a clutch may be higher for a young goose than for older individuals that have gained more experience, and the benefits derived from breeding early may not outweigh the costs. In this section we explore costs and benefits of different life-history trajectories.

Environmental conditions experienced in different Svalbard breeding areas explained 4.5% of the variation in lifetime reproductive success (Figure 9.10). This effect may be due to differences in habitat type and quality, predation pressure, or to differences in density of the local goose population. In Chapter 14 these issues are explored further. Another 10.9% of the variation in lifetime success was due to variation among cohorts, i.e. among the years included in the birds' lifetimes. Whereas location and birth year accounted for some of the variation in lifetime reproduction, the remaining 84% of variation can be attributed to individual characteristics, which are discussed below.

Individual qualities

Using records from one of our best-studied Svalbard cohorts, only 55% of the females (n = 196) and 48% of the males (n = 175) returned to the wintering grounds with one or more offspring during their lifetimes (Figure 9.11). The females that did reproduce had an average of 4.3 (SE 0.3) offspring in a lifetime. The most successful goose recruited offspring to the wintering grounds in eight of her 20 years of life, resulting in a grand total of 21 goslings that survived through autumn. Figure 9.11 also indicates the extent of variation that was explained by the three components of lifetime reproduction. The most influential parameter

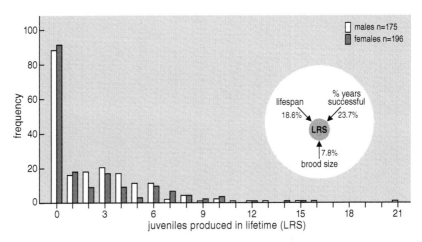

Figure 9.11. *Frequency distribution of lifetime reproductive success in Barnacle Geese, and percentages of variance explained by breeding parameters. Only half of the population reproduced successfully during their lifetimes. The most successful female and male produced 21 and 16 offspring, respectively. The overall average, including individuals that never bred, was 2.4 goslings (SE 0.2) for females and 2.1 goslings (SE 0.2) for males. Reproductive success was determined by counting number of mature goslings (> four months of age) associating with marked adults after arrival on the Scottish wintering grounds. Analysis limited to the 1976 cohort captured as yearlings on Nordenskiöldkysten, thus excluding birds that died in the first year. Updated from Owen & Black (1989b, 1991b).*

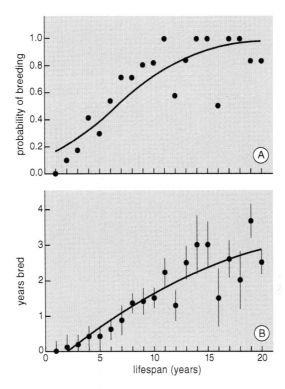

Figure 9.12. *Relationship between breeding and lifespan in Barnacle Geese. (A) Probability of raising at least one gosling surviving through autumn migration to Scotland (i.e. ≥ four months of age) in relation to lifespan. (B) Number of successful breeding years in relation to lifespan (y = −0.45 + 0.24x − 0.004x²). Plots show an increased breeding success with lifespan. Slopes for males were similar.*

was the proportion of successful years (i.e. with goslings in winter), followed by lifespan and brood size. While the probability of successful reproduction increased with lifespan (Figure 9.12),[84] a long life did not guarantee success; 11 females and 18 males in this cohort that lived for 10 or more years still failed to reproduce.

The highest lifetime success was achieved by individuals that bred successfully in the early years of a long life (individuals in the right bottom corner of Figure 9.13A). Individuals that bred successfully only at an older age did not perform as well. Most geese became successful at a later age than the most successful trajectories, and lived a shorter life (compare Figure 9.13A,B). This means that relatively few individuals achieved high levels of success and these few produced the majority of offspring. The most successful females producing 50% of all the goslings represented only 13% of the cohort.

A potential mechanism for the individual variation in reproductive success is referred to as the *social feedback loop* (Lamprecht 1986b, 1990). This hypothesis states that successful geese are more likely to be successful again because of their improved dominance status and access to food that come from maintaining a family unit in the winter (Black & Owen 1989a). Once a goose experiences success and continues association with mature offspring, its access to food may improve due to enhanced dominance status, thus increasing the probability of survival and further reproductive success (Black 2001). It may be that these

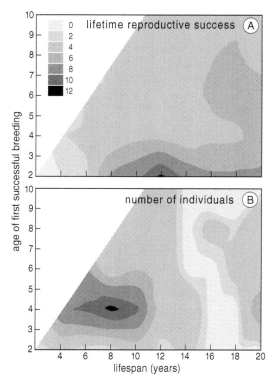

Figure 9.13. *(A) Contour plots of lifetime reproductive success, and (B) the number of individuals in the landscape of lifespan (x-axis) and age of first successful breeding (y-axis) as assessed during winter brood size observations in Scotland. The plots show that while most individuals bred successfully when four years old, the most productive geese bred at an earlier age, and lived longer. Not included in the figure are individuals that never bred successfully. The analysis includes females of the 1976 cohort from Nordenskiöldkysten.*

high-quality birds are better able to withstand the stresses and strains of reproduction than low-quality birds. In Chapter 10 we discuss the significance of body size (skeletal size) on reproductive performance, where larger birds are capable of carrying larger nutrient stores for use at critical moments in the annual cycle.

Unanswered questions

Additional investigations are needed to understand the cumulative stresses involved in goose reproduction. These stresses eventually take their toll, and cause the pattern of old age senescence and the reduced likelihood of survival after successful breeding episodes. Compared to pairs without offspring, parent geese feed less because they are more vigilant and spend more time defending feeding space (Lazarus & Inglis 1978, Prop *et al.* 1980, Lessells 1987, Gregoire & Ankney 1990, Sedinger & Raveling 1990), and parents of larger broods spend more time being vigilant and competing with flock members (Lessells 1986. Schindler & Lamprecht 1987, Black & Owen 1989a,b, Forslund 1993, Sedinger *et al.* 1995, Loonen *et al.* 1999, Siriwardena & Black 1999). This parental burden may continue for almost a full year if goslings remain in the family (Chapter 6). Reduced feeding time may impair parents' body condition, and thus individuals may be less well prepared for adverse conditions. Likewise, parents might experience elevated levels of androgens and corticosteroids that eventually take their toll on immune defence against disease and parasites (Gustafsson *et al.* 1994, Deerenberg *et al.* 1997, Norris & Evans 2000, Hanssen *et al.* 2003). Exploring the physiological mechanisms involved in sustained and recurring parental investment, whether deterioration in physical ability or degradation of the immune system, would enable better understanding of waterfowl life histories (*sensu* Ricklefs 2000, Zera & Harshman 2001, Bize *et al.* 2004). Measuring androgens and corticosteroids obtained from goose droppings is a useful approach for exploring some of these mechanisms (Kotrschal *et al.* 1998, Hirschenhauser *et al.* 1999, Scheiber *et al.* 2013).

Summary and conclusions

The average annual survival rate for most Barnacle Goose age classes was about 0.90, and this decreased to less than 0.70 for individuals of 20 years and over. Most mortality of adult geese was realised during migration between the breeding and wintering grounds. Adult female survival was related to breeding status; fewer successful breeders survived to the next year than failed or non-breeders. Males survived at about the same rate regardless of their reproductive status, indicating that the roles performed in reproduction were substantially different between the sexes. Females with deteriorating body condition during incubation spent more foraging time off the nest, which increased the risk of nest failure. Only abandoning the clutch, and proceeding with the moult early in the season, allowed sufficient time for recuperation of body condition to avoid a reduction in survival in late summer and during autumn migration. Barnacle Goose females generally sacrificed current breeding attempts in favour of future reproduction. Though the majority of individuals attempted to breed each year by establishing a nest and laying and incubating eggs, relatively few actually succeeded in rearing offspring that survived to return to the wintering area; 80% established nests and only 17% returned with goslings. Population density also influenced reproduction. In a comparison of cohorts living in the same breeding area during a time of increasing population densities, we showed that a higher proportion of the earlier cohort (with a lower density) bred more successfully (with offspring in winter) than later cohorts. About half of the population (45% of the females and 52% of the males) never successfully reproduced during their lifetimes.

The average female, including the unsuccessful breeders, produced 2.4 (SE 0.2) offspring during a 9.5-year lifetime. Lifetime reproductive success is best explained by the number of actual years of success, more so than the birds' lifespan or the number of goslings in each brood. The majority of juveniles were produced by just a minority of very successful breeders; 13% of the birds which began to breed successfully at an early age (two to four years) and lived to an old age (10+ years) were responsible for producing 50% of all the surviving offspring. The most successful bird produced offspring in eight of her 20 years resulting in 21 surviving goslings, indicating that successful breeding was intermittent.

Statistical analyses

[77] An assessment of Barnacle Goose annual survival by age was made using resighting records of 44 Barnacle Goose females and 46 males from the 1972 cohort from Dunøyane colony (Svalbard) (Figure 9.1). See Appendix 1, Table 6 for details of selected models.

[78] An assessment of annual survival (by year) was made using resighting records of 3,428 Barnacle Goose females and 3,429 males in relation to breeding status (Figure 9.2). See Appendix 1, Table 7a,b for details of selected models.

[79] ANOVA: Time off the nest to forage was negatively related to body condition as measured by Abdominal Profile Index ($F_{1,136} = 31.9$, $P < 0.0005$; Figure 9.4A). ANOVA: The slope was steeper in years with a late snow melt ($F_{1,136} = 3.96$, $P < 0.05$). Logistic regression: Proportion of females successfully hatching a clutch in relation to average length of incubation breaks ($\chi^2 = 15.25$, df = 1, n = 77, $P < 0.0005$; Figure 9.4B); the earliness of the snowmelt did not have an additional effect ($\chi^2 = 1.68$, NS). Each female features only once in the analysis.

[80] Kruskal–Wallis test: Residual vigilance among three experimental treatments, including reduced, control and enlarged broods ($\chi^2 = 17.10$, $P = 0.001$; Figure 9.5). Residual vigilance was calculated after correcting for sex, hatch date and days since hatch. From Loonen *et al.* (1999).

[81] ANOVA: Body mass of adult females in relation to brood size ($F_{1,27} = 4.88$, $P = 0.036$ with a slope of 33g per gosling; Figure 9.6). Kruskal–Wallis test: Average residual mass of adult females in relation to experimental categories (reduced, control, enlarged broods; approached significance, $\chi^2 = 5.56$, $P = 0.06$). Broods were manipulated soon after hatch. From Loonen *et al.* (1999).

[82] Logistic regression: Proportion of female Barnacle Geese bringing at least one gosling to the Scottish wintering area in relation to age ($\chi^2 = 199.5$, df = 2, n = 3,047, $P < 0.0001$; Figure 9.8). A similar relationship was described for males. From Black & Owen (1995).

[83] Proportion of female Barnacle Geese bringing at least one gosling to Scotland for three different Svalbard cohorts across years (Figure 9.9). The analysis shows trends over the years that differ among cohorts ($F_{2,3041} = 4.33$, $P = 0.013$); regression slopes for the cohorts: 0.003, –0.005 and –0.010, respectively. Observations refer to three cohorts (1976, 1980, 1986) from the Nordenskiöldkysten brood-rearing area. Reproductive success was corrected for age.

[84] Logistic regression: Probability of raising at least one gosling surviving through autumn migration to Scotland (i.e. ≥ four months of age) in relation to lifespan ($y = -3.64 + 0.77x - 0.025x2$ on a logit scale; $\chi^2 = 78.6$, df = 2, $P < 0.0001$, n = 196; Figure 9.12). GLM with Poisson distribution: Number of successful breeding years in relation to lifespan ($y = -2.82 + 0.47x - 0.014x2$ on a logit scale; $\chi^2 = 156.5$, df = 2, $P < 0.0001$). Data for males are not presented but the relationships were similar.

CHAPTER 10
Body size

Body size measurements are important in wildlife studies because they may provide information about individuals' genetic make up and early development. In geese, structural traits (i.e. bones) grow at different rates, but most reach final size within the first year. Some bones grow quite quickly; for example, at fledging tarsus lengths of goslings are already close to full size, whereas the head reaches full size somewhat later (Larsson & Forslund 1992, Loonen *et al.* 1997b). Once body structures have matured they will not show further growth even if foraging opportunities and other environmental conditions improve. Consequently, a small head or short tarsus might indicate that the individual experienced poor conditions during the first months of life.

In this chapter we describe factors that affect the final body size of geese and make comparisons between the Svalbard and Baltic populations. We show that egg size and final body size in Barnacle Geese are influenced by a complex interaction between genes, environmental conditions and individual decisions. We provide evidence that structurally large birds, on average, produced more goslings than smaller birds, but also that average body size decreased over time when numbers and grazing intensity increased in the study colonies. We begin by quantifying the effects of genetic and environmental factors on body size.

Body size heritability

One method to statistically quantify genetic and environmental effects on body size is to compare traits of a large set of parents and offspring. If body size traits of offspring strongly resemble those of their parents we may argue that individual variation to a large extent

173

reflects underlying genetic variation. In such cases, the resemblance is assumed to be a consequence of the expression of genes that parents and their offspring share (Box 10.1). However, when interpreting resemblance analyses one should recognise that confounding non-genetic mechanisms may exist, such as common environment effects and maternal effects, which may also influence resemblance between relatives. Other methods to quantify and separate genetic and environmental effects on body size include, for example, cross-fostering experiments and analyses of body size variation among different year-classes of offspring with the same or a similar set of parents.

In the Baltic population of Barnacle Geese we were able to follow and measure a large set of related individuals (Larsson & Forslund 1992). Our initial analyses of the resemblance

BOX 10.1 QUANTIFYING BODY SIZE VARIATION AND ESTIMATING HERITABILITY

In this chapter, we use head and tarsus length measurements as indices of structural body size. In some cases it is also valuable to combine several measurements into one single size variable. By using a Principal Component Analysis (PCA) one can often capture much of the overall size information into a single virtual size variable – the first principal component (PC1). Use of body mass, after controlling for structural body size, is often reserved for describing a bird's condition.[85]

Heritability is the proportion of phenotypic variation in a population that is due to genetic variation, or more specifically to additive genetic variation. Heritability estimates can be obtained by comparing phenotypic traits, like body size measures, of large numbers of parents and offspring (Falconer & Mackay 1996). Estimates of heritability (h^2) may vary between zero and one, where h^2 refers to a ratio between the additive genetic variance (i.e. variation caused by the additive effects of many genes) and phenotypic variance (i.e. the directly observable variation in a trait, like head and tarsus lengths). If a trait has a heritability of, for example, 0.40, it may respond quickly (as a microevolutionary response) to new selection pressures because there is a large amount of additive genetic variation for selection to act on in the population. However, even in such a case, 60% of the phenotypic variation can be explained by environmental factors.

Heritability estimates can be obtained by several methods, which have different advantages and disadvantages. One common method is to perform linear regressions of mid-offspring values (i.e. average values of offspring) on mid-parent values (i.e. average values of the father and mother) or on single-parent values (i.e. values of mothers or fathers). The slope of the regression may then be used as an estimate of heritability. When single-parent values are used in linear regression analyses, heritability is calculated as twice the slope of the regression line (Falconer & Mackay 1996). Other methods to estimate heritability rely on analyses of variation within and among broods of half or full siblings. By comparing the results from different methods one may elucidate different genetic effects, environmental effects, and various interactions between them.

Estimates of genetic correlations between body size traits can also be obtained by comparing trait values of related individuals. Positive genetic correlations between body size traits may be explained by pleiotropic effects of genes; that is, the same genes may affect several traits at the same time (Falconer & Mackay 1996). The presence of genetic correlations may lead to a situation where selection on one trait affects the evolutionary response of other traits.

between fully grown offspring and parents (using average values of both parents), and among siblings, revealed that structural body size traits, including head and tarsus length as well as egg size, were heritable to a substantial degree. The heritability estimates (h^2) varied between 0.38 and 0.67 (Table 10.1). Body size traits were also genetically correlated, indicating that the same genes may affect several traits.

Further in-depth analyses also revealed that, on average, the tarsus length of offspring (a trait that usually reaches full size before fledging) resembled that of their mothers more than that of their fathers. The heritability estimates for tarsus length obtained from offspring–mother regressions ($h^2 = 0.80$) were considerably higher than estimates from offspring–father regressions ($h^2 = 0.18$) (Figure 10.1). Similar results have been found in other bird studies and have sometimes been explained as an effect of extra-pair fertilisations where some offspring are fathered by males other than those actually observed guarding or feeding them; that is, by males other than those that were actually classified as fathers in the analyses. Molecular genetic evidence suggests, however, that extra-pair fertilisations are rare in Barnacle Goose populations (Chapter 7). Other mechanisms must, therefore, be invoked

Table 10.1. *Heritability estimates (h^2) for body size and egg size in the Baltic population of Barnacle Geese. Estimates were obtained from regressions of measurements of offspring on mid-parents. Genetic correlations (r_A) between body size traits are shown to the right. Sample sizes (N) = number of families. All estimates were significantly different from zero. Updated from Larsson & Forslund (1992).*

Trait	N	h^2	SE		N	r_A	SE
Head length (HL)	218	0.38	0.07	HL *versus* TL	105	0.55	0.13
Tarsus length (TL)	218	0.39	0.09	TL *versus* BM	105	0.88	0.03
Body mass (BM)	218	0.54	0.07	HL *versus* BM	105	0.77	0.07
Egg size[a]	37	0.67	0.27				

[a] h^2 estimate obtained from daughter–mother regression. Egg size (volume) was calculated from a regression equation based on length and width of the egg (Volume = $0.4776 \times$ egg length $\times$ (egg width)$^2 + 6.462$).

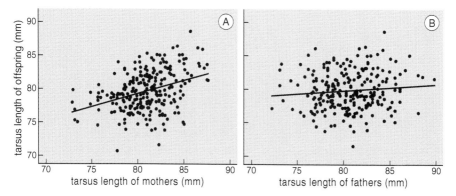

Figure 10.1. *Resemblance between tarsus lengths of offspring and (A) mothers and (B) fathers. The heritability estimate obtained from offspring–mother regression ($h^2 = 0.80$, SE = 0.12, n = 253 families) was considerably higher than the estimate obtained from offspring–father regression ($h^2 = 0.18$, SE = 0.12, n = 244 families).*

Sex	N	Natal grazing site	Preferred grazing site when with own offspring		χ^2	P
			A	B		
Females	39	A	25	14		
	32	B	2	30	24.96	< 0.0001
Males	30	A	14	16		
	16	B	4	12	2.06	NS

Table 10.2. *Tendency in Baltic Barnacle Goose females and males to return to natal grazing sites when leading their own broods. Determined for 117 birds whose grazing sites were known when they led broods and when they were goslings being led by their parents. Based on an average of four observations per family. Analysis shows there was a strong tendency for females to return to their natal grazing site when leading young. From Larsson & Forslund (1992).*

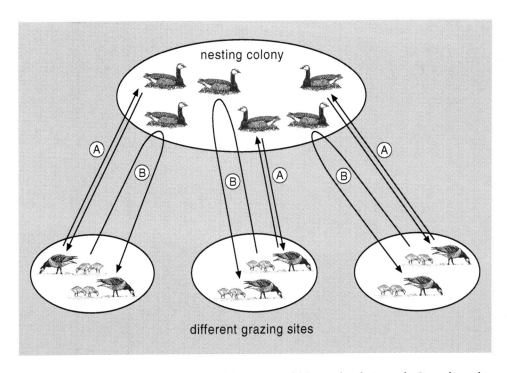

Figure 10.2. *Illustration of socially inherited foraging site fidelity in female Barnacle Geese observed in the Baltic population. Breeding pairs in the largest nesting colony use a variety of grazing sites during brood rearing. Grazing sites are of different quality, which affects the growth and final body size of goslings. Breeding pairs are usually faithful to brood-rearing sites and use the same sites year after year (A). Females who were reared at certain sites often lead their goslings to those same sites when breeding several years later (B). This behaviour created a common environment effect between mothers and offspring, resulting in an enhanced resemblance because they had been exposed to similar environmental conditions during growth. Drawing by Kjell Larsson.*

to explain the stronger resemblance of offspring to mothers than to fathers. We uncovered a clue when we studied the brood-rearing behaviour of marked individuals and found that female parents led their newly hatched goslings to brood-rearing sites where they had been led as goslings several years earlier (Table 10.2). Such site fidelity was much less pronounced in males. This phenomenon of socially inherited foraging site fidelity in females is illustrated in Figure 10.2. Because vegetation at different brood-rearing sites differed in quality for growing goslings, a *common environment effect* between mothers and offspring appeared (*sensu* Falconer & Mackay 1996). Thus, the enhanced resemblance between mothers and offspring was to some degree due to the fact that they had often grown up at the same brood-rearing sites.

In summary, it is clear that body size variation among individuals to a substantial degree reflects underlying genetic variation. Body size may therefore respond quite quickly if new selection pressures arise. However, it is also clear that non-genetic mechanisms, including decisions made by parents about where to lead their goslings, can to some extent influence the resemblance between related individuals and inflate heritability estimates for early maturing traits (Larsson & Forslund 1992, Larsson 1993). In subsequent sections we continue to explore important links between early environmental conditions, food supply and final body size.

Food supply shapes body size

Environmental factors, including weather, microclimate and nutrients, have a large impact on availability of high-quality goose food. Sometimes precipitation is a good proxy to food quality. In contrast to waterlogged tundra habitats in the Arctic, Baltic habitats often suffer from low precipitation and high evaporation in spring, and therefore the amount of rain is bound to influence goose food. In years with large amounts of rain in May and June the

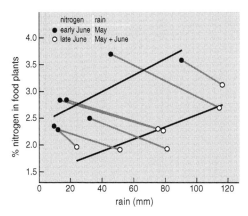

Figure 10.3. *Quality of a preferred food plant (*Festuca rubra*) of grazing Baltic Barnacle Geese in relation to the amount of rain during brood rearing for seven years. Filled and open symbols show the nitrogen content at the beginning of June (around one week after average hatching date) in relation to the amount of rain in May and nitrogen content at the end of June in relation to the total amount of rain in May and June, respectively. Nitrogen content of the food plants was higher in years with much rain. Data from the same years (filled and open symbols) are connected, showing that the nitrogen content of the food decreased over the brood-rearing period in all years. Each point denotes the average nitrogen content of grass in three to five plots within an important brood-rearing area. From Larsson* et al. *(1998).*

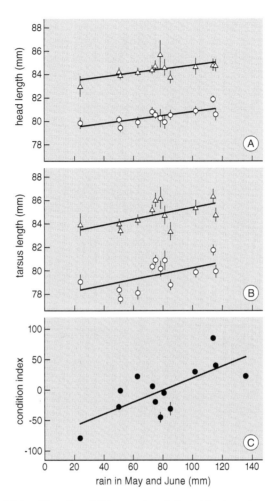

Figure 10.4. *Average (A) head length and (B) tarsus length from 12 cohorts born and reared in the main Baltic colony in relation to the amount of rain in May and June in the year of birth. Males indicated by triangles, females by circles. Sample sizes: range 28–129, average 58 birds. (C) Yearly condition indices of adults in relation to the amount of rain in May and June. Condition indices of adults were calculated from body mass measurements using a GLM model.[85] Error bars denote SE. From Larsson* et al. *(1998).*

nitrogen content of food plants was higher during the gosling growth period in June (Figure 10.3).[86] The close correlation between availability of high-quality food and the condition of adults and average final body size of different cohorts suggests a strong causal relationship (Figure 10.4).[87] Furthermore, birds that were born late in the season achieved, on average, a smaller final size than birds born earlier, which we also attribute to within-season shifts in food quality.[88] A delay in hatching date of only 10 days resulted in 2.3mm (2.7%) shorter tarsi at adult age (Larsson & Forslund 1991).

We also found strong indications that growth of goslings and final size were affected by the local density of birds (Black *et al.* 1998, Larsson *et al.* 1998). At a Svalbard colony, Loonen *et al.* (1998) showed that the same parents produced smaller offspring in situations with increased goose density. The presence of Arctic Foxes during brood rearing reduced the foraging area that could be exploited by broods. In those years, gosling growth rate was substantially less than in years without foxes and when goose broods used much larger feeding areas.

Taken together, these findings show that environmental effects to a significant degree influence final structural size. In some cases differences in average adult body size between cohorts were larger than body size differences observed between the sexes. Additional environmental factors have also been linked with growth and average body size of cohorts in other goose populations (Cooch *et al.* 1991a,b, Sedinger & Flint 1991, Lindholm *et al.* 1994, Fondell *et al.* 2011). The average body size of a sample of individuals reared at a specific goose rearing area can therefore be used as an indicator for the environmental conditions they experienced during early life.

Body size affects reproductive ability

The body size of a bird may directly or indirectly influence different behaviours and abilities. In Chapter 4 we argued that body size may influence the mate choice process and competitive situations, but what about reproductive ability?

Analyses of more than 2,000 ringed individuals attempting to breed in the main Baltic study colony during a 12-year period showed that structurally large females generally

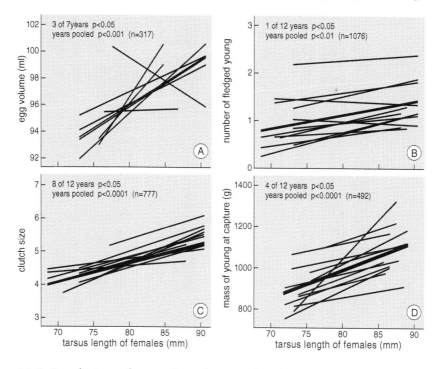

Figure 10.5. *Reproductive performance (egg volume, number of fledged young, clutch size and mass of young at capture) in relation to tarsus length as an index of body size in female Barnacle Geese from the primary Baltic study colony between 1984 and 1996. Thin lines represent linear regressions of data obtained in different years. Thick lines represent linear regressions obtained from pooled data. In the pooled regressions a specific individual occurred only once in the analyses. From Larsson* et al. *(1998).*

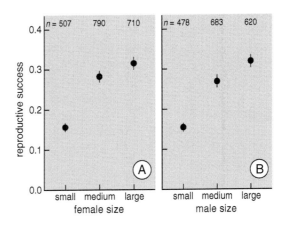

Figure 10.6. *Average reproductive success (number of goslings with marked birds on wintering grounds) for different sized (A) female and (B) male Barnacle Geese in the Svalbard population. Sample sizes are indicated. Analysis controlled for the effects of year, age and breeding area. From Choudhury* et al. *(1996).*

produced larger eggs and larger clutches, and more and heavier young, than smaller females (Figure 10.5).[89] Similar relationships were also observed in the Svalbard population, where the average brood size in winter was positively related to the structural body size of parents (Figure 10.6).[90] Furthermore, in our best-studied 1976 cohort from Nordenskiöldkysten, body size was positively correlated with each of the most important components of lifetime reproductive success; in order of importance: lifespan, number of years successful, and brood size.[91] Figure 10.7 shows that the largest females lived for about 9.5 years, bred successfully 1.5 times and produced three surviving goslings in their lifetimes. In contrast, small females lived for seven years, bred successfully 0.7 times and produced only 1.5 offspring. In contrast to the situation in the Svalbard population, where lifespan and body size were correlated in females, the survival rate of adults in the Baltic study population was not related to structural body size (Larsson *et al.* 1998).

Although data presented are correlative (i.e. evidence for a causal pathway is not explored), the above findings indicate a clear advantage to large structural size. The results are in accordance with the hypothesis that structurally larger bodies are capable of transporting more fat and nutrients from winter and spring staging areas to breeding habitats than smaller bodies, and that the extra resources allow larger females to produce larger clutches and broods. A similar mechanism might also be important for males; individuals with larger nutrient stores might be able to invest more in nest defence during incubation and gosling defence after hatch. Positive relationships between body size and reproductive ability have also been found in Snow Geese and Black Brant (Ankney & MacInnes 1978, Alisauskas & Ankney 1990, Sedinger *et al.* 1995; but see contrary results from Cooke *et al.* 1990, 1995, Cooch *et al.* 1992).

Even though we found that larger individuals generally performed better than smaller individuals we also found evidence that the size of a bird's partner mattered. For example, in some years we found that small females had higher reproductive success when paired with a small male than with a large male (Choudhury *et al.* 1996). It is therefore likely that pair-size combinations and compatibility also affect reproductive success (Chapter 4).

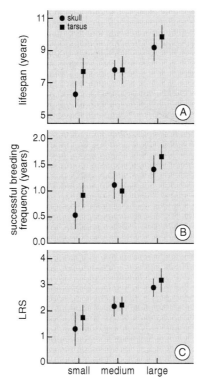

Figure 10.7. *Average lifespan, number of successful breeding years, and lifetime reproductive success (LRS) in relation to female body size (skull and tarsus length). Successful breeding was determined by counting mature goslings on the Scottish wintering grounds. (A) Lifespan – larger females lived longer than smaller females. (B) Number (frequency) of successful breeding years – larger females bred more often in their lifetimes than smaller females. (C) Lifetime reproductive success – larger females produced more offspring during their lifetimes than smaller females. Sample sizes 25–85 for each point. Average ± SE are indicated.*

Body size changes in populations over time

We have shown that structural body size traits varied considerably among individuals and were heritable. Thus, if there is selection in action, for example if larger birds are consistently more successful in reproduction or enjoy higher survival rates than smaller birds, there is potential for relatively rapid evolutionary changes. In our study populations, therefore, we should have expected the average body size to increase. However, what we observed was an opposite trend at several study colonies. To elucidate the seeming contradiction we continue by pointing to the effects of increasing population densities on growth and final body size in the study populations.

On Svalbard, the body size of cohorts reared at Kongsfjorden declined substantially over time, whereas the decline was less pronounced at Nordenskiöldkysten, apparently reflecting a difference in population density in these areas (Figure 10.8).[92] Both colonies increased in number but the Kongsfjorden colony experienced a much faster growth rate (12% per year) than the Nordenskiöldkysten colony (5% per year). The apparent link between population

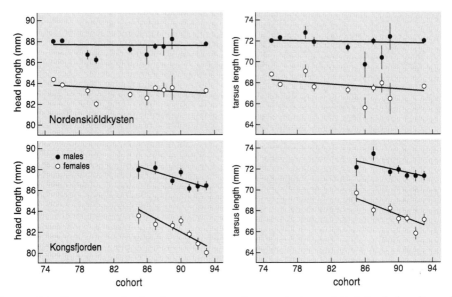

Figure 10.8. *Variation in head and tarsus measures for several cohorts of birds hatched and reared at two Svalbard colonies. There was a highly significant long-term decline in body size at both locations. The decline at Kongsfjorden was steeper. From Black* et al. *(1998).*

growth rate and the decline in body size corresponds well with the *habitat saturation hypothesis*, stating that body size reduction is related to the amount and quality of food available per gosling during the initial period of skeletal growth before fledging. In the main Baltic study colony the observed body size decline was most likely due to a similar density-dependent process which was mainly in effect during the very early phase of colony growth (Larsson *et al.* 1998).

We cannot exclude that during other circumstances or time periods opposite selection pressures on body size may be more important. Large body size may pose problems for flight manoeuvrability or long-distance migrations (Pennycuick 1989). Hence, smaller geese may be better able to avoid aerial attack from predators and the costs of migration might be less than for larger geese. Furthermore, in situations when the overall availability of high-quality food is limited during brood rearing, as for example in very dry summers in the Baltic, goslings produced by small parents might be at an advantage because they may require smaller absolute amounts of food to survive. We conclude, however, that during our studies in the Baltic and on Svalbard possible upward microevolutionary responses to selection were obscured by a stronger density-dependent food limitation in brood-rearing areas.

Summary and conclusions

Body size varied among individuals, among cohorts within colonies and among colonies. While some of this variation was due to genetic variation, we found that a substantial amount was also the result of environmental conditions that goslings experienced during the brood-rearing period. In the Baltic, precipitation was found to affect the nutritive value of foods, which in turn influenced the condition, growth and final body size of the geese. We provide evidence showing that females leading broods often return to the foraging areas that their parents took them to several years earlier. The body size of a bird was therefore the result of the combined effects of its genes, the environmental conditions during growth, and the decisions made by the bird itself and its parents.

Body size affects reproductive abilities. For example, body size correlated positively with several measures of reproductive success and survival. In the Baltic population, larger structural size enabled the production of larger eggs and larger clutches, and resulted in more and larger goslings that fledged successfully. In the Svalbard population, reproductive success (brood size in winter) was positively correlated with body size for both females and males. However, we showed that in spite of strong selection for larger body sizes (i.e. that larger birds produce more young), average body sizes of cohorts were actually declining as colonies become larger. We favour the *habitat saturation hypothesis* as an explanation for declining body sizes, which suggests that increasing competition for limited food results in reduced food intake rates for goslings when skeletal growth occurs in the first weeks after hatch. It seems that an upward response to selection for larger body sizes was obscured by a large environmental effect; that is, by density-dependent food limitation in brood-rearing areas.

Statistical analyses

[85] Sources for effects that we controlled for in the GLM model included PC1, date of capture (continuous variables), sex and year (class variables). Date of capture was included in the model because body mass generally decreases over the moulting period. Least square means (i.e. adjusted average body mass values) were computed for all years. Yearly condition indices were then calculated by subtracting the overall average from the year-specific least-square means. From Larsson *et al.* (1998).

[86] Both regression lines in Figure 10.3 were significantly positive (b = 0.015 and 0.012, respectively, n = 7, P < 0.05). From Larsson *et al.* (1998).

[87] ANCOVA head length: rain, $F_{1,21}$ = 18.8, P < 0.001; sex, $F_{1,21}$ = 400.4, P < 0.0001 (Figure 10.4A); tarsus: rain, $F_{1,21}$ = 10.4, P < 0.01; sex, $F_{1,21}$ = 168.4, P < 0.0001 (Figure 10.4B). Yearly condition indices of adults in relation to the amount of rain in May and June (b = 0.98, r^2 = 0.52, n = 13, P < 0.01 (Figure 10.4C)). From Larsson *et al.* (1998).

[88] Final tarsus length (n = 401 birds) was a function of relative date of hatch (a continuous variable) (F_1 = 11.78, P < 0.001) after controlling for year of birth, grazing site used before fledging, and sex. The regression coefficient for tarsus length on relative date of hatch was negative and significantly different from zero (b = −0.23, P < 0.001). From Larsson & Forslund (1991).

[89] Average egg size (volume) was calculated for the entire clutch. Egg size was primarily determined by the female and not the male parent (Larsson & Forslund 1992). Clutch size in nests of marked pairs was recorded at the middle or end of the incubation period in May. Number of fledged young was determined at one to two weeks prior to fledging. Body mass of captured young was measured with a Pesola spring balance to the nearest 25g.

[90] Average reproductive success (number of goslings with marked birds on wintering grounds) for different sized (A) female (GLIM χ^2 = 12.8, df = 2, P < 0.005),and (B) male (GLIM χ^2 = 10.3, df = 2, P < 0.01) Barnacle Geese in the Svalbard population. Size was based on an index of skull and tarsus using the first principal component (PC1). See Figure 10.6. From Choudhury *et al.* (1996).

[91] GLM Lifespan: skull lengths $F_{1,188}$ = 5.34, P = 0.022; tarsus lengths $F_{1,187}$ = 8.24, P < 0.005 (Figure 10.7A). Number of successful breeding years: skull lengths $F_{1,188}$ = 7.75, P < 0.006; tarsus length $F_{1,187}$ = 7.81, P < 0.006 (described as average ± SE frequency of years in Figure 10.7B). Lifetime reproductive success: skull $F_{1,188}$ = 6.17, P = 0.014; tarsus $F_{1,187}$ = 5.66, P = 0.018 (Figure 10.7C). The relationship between structural size and successful breeding years and lifetime reproduction was also apparent for males.

[92] ANCOVA Nordenskiöldkysten: head length $F_{1,2179}$ = 13.76, P < 0.001; tarsus length $F_{1,2179}$ = 6.83, P = 0.009; Kongsfjorden: head length $F_{1,635}$ = 34.92, P < 0.001; tarsus length $F_{1,635}$ = 18.42, P = 0.001. Sex was included as a classification factor. Birth year (or cohort) of birds marked as goslings and yearlings was determined during capture sessions. For this study site we assumed birds caught as yearlings were hatched and reared in the area of capture. Pooling the sexes for both colonies, there was a highly significant difference in body size variation between colonies during 1983–1993 (MANCOVA $F_{2,1443}$ = 16.80, P < 0.001). From Black *et al.* (1998). Note that tarsus was measured in a different way in the Baltic and Svalbard. Absolute measures in Figures 10.4 and 10.8 are therefore not directly comparable.

CHAPTER 11
Timing

Animals dedicate much effort to preparing themselves for energy-draining activities. Like other Arctic-breeding geese, Barnacle Geese accumulate body stores in the form of fats and proteins that can be used later in the year. Favourable conditions for deposition of stores are relatively rare, and preparations must take place during peaks in food availability. For much of the year energy acquisition is constrained by environmental conditions; during cold, short winter days, for example, geese may not be able to acquire enough food even to maintain a constant body mass. Therefore, geese must time the events in their annual cycle to take advantage of foraging opportunities. This chapter explores the energetic constraints in the annual cycle and the influence that timing of migration and egg-laying has on reproductive success.

To discuss the problem of properly timing energy-demanding events we introduce the term *productive energy*. Productive energy is the daily amount of energy metabolised above maintenance needs, and represents the amount of energy that can be invested in body growth, body stores, or reproduction. Figure 11.1 depicts the annual cycle of some key parameters relevant to productive energy for Barnacle Geese, in this case for the population breeding in Svalbard. Since foraging is primarily restricted to daylight hours, the first variable of interest is the period of daylight experienced through the year (Figure 11.1A). Of particular importance is the possibility of unrestricted feeding in the Arctic light regime. Upon arrival in Svalbard geese benefit from permanent daylight by spending a considerable proportion of their time feeding. The quality of food on offer varies over the year (using protein content as a proxy, Figure 11.1B). The elevated protein values in the wintering area reflect management practices with extremely high fertilisation rates, as are common in present-day farming (Chapter 15). Food quality is low at arrival on the breeding grounds, when mosses together

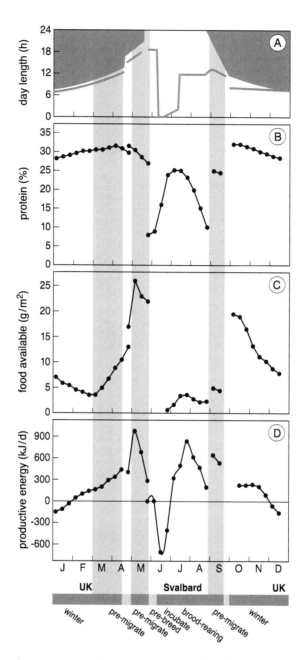

Figure 11.1. *Seasonal variation in productive energy generated by Barnacle Geese. Bottom axis indicates successive periods in the annual cycle. Three pre-migratory periods are indicated by shading; early spring at the Solway Firth (UK), spring along the Norwegian coast, and late summer in Svalbard. Productive energy is associated with (A) day length, and time spent foraging (indicated by grey lines in the light period of day); (B) food quality; and (C) food abundance (edible biomass of main food plants). (D) Daily productive energy calculated as the increment of energy metabolised above maintenance needs. Calculations of productive energy are based on faecal output per day combined with digestibility of food, for 10-day periods (details in Prop & Black 1998, Chapter 3). From Prop (2004).*

with plant remains from the previous year are the main food items. When fresh food plants appear after snow recedes from the tundra (Chapter 2), food quality rapidly improves. With ageing of the plants there is a steep decline in quality in the course of summer. The trend in food biomass (Figure 11.1C) demonstrates the latitudinal gradient in commencement of spring growth, initiated at the end February at the Solway Firth (UK), the end of April in Norway, and the beginning of July on the Arctic breeding grounds (Svalbard).

Productive energy is shown here for adult female Barnacle Geese (Figure 11.1D). During midwinter in Scotland there may be a period of energy deficit (Owen *et al.* 1992), when the geese fall back on their body stores. With advancing plant growth in spring, progressively larger stores are accumulated before the 1,500km migration to Norway. The Norwegian staging areas provide the highest daily gains of the entire annual cycle. With the stores accumulated in Norway the geese proceed an additional 1,500km to Svalbard. Barnacle Geese spend on average two to three weeks in Svalbard before egg-laying. At this time of the year conditions to find food are severe, in particular when fresh snow or temperatures below freezing hamper feeding. The rapid increase of the abdominal profile by geese feeding close to bird cliffs (Hübner 2006, Hübner *et al.* 2010), though, suggests that the geese are in positive energy balance and that they are able to accumulate energy and nutrients needed for reproduction. The importance of this pre-breeding period in Svalbard was supported by Hahn *et al.* (2011), who were able to distinguish the origin of egg contents (yolk and albumin) based on assessments of the ratios of stable isotopes. No less than *c.* 50% of the egg contents were derived from nutrients collected in Svalbard (Hahn *et al.* 2011).

Laying and incubation follow, a period of near starvation for the breeding female (Tombre & Erikstad 1996). For geese that time their breeding effort appropriately, eggs hatch as nutritious food plants begin to emerge in brood-rearing habitats, enabling body condition of breeding birds to recover and goslings to grow. The parents undergo moult at this time and regain flight approximately when the goslings take wing at the end of August. They then begin to visit the vast, almost unexploited tundra and the lush vegetation on the slopes under seabird cliffs. This results in a fourth peak of productive energy, which fuels the autumn migration. Once the geese are back in the UK, the vegetation allows a further 30-day period of positive energy balance before the onset of colder weather and short days.

In the following sections we describe in more detail the limitations on the ability of Barnacle Geese to acquire sufficient energy to fuel seasonal requirements. Subsequently we discuss a central question: what are the fitness consequences associated with the timing of migration and egg-laying? Finally, the question of individual decisions within this framework is addressed.

Digestive bottleneck

Three key parameters determine the rate of energy acquisition by geese: the concentration of metabolisable energy of their food (ME, kJ/g), ingestion rate (IR, g/min foraging) and digestive capacity (g/min). Metabolisable energy refers to energy that is assimilated after food is broken down during the digestive process. To metabolise sufficient energy geese need to select food that allows an appropriate combination of ME and IR. A low food quality can be compensated for by a high ingestion rate, and vice versa. Optimal combinations are

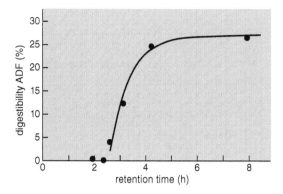

Figure 11.2. *Relationship between digestibility of acid detergent fibre (ADF, an important component of cell walls) and food retention time. Data points represent average values per seasonal period. Digestibility was determined by using lignin as a marker, and retention times were based on intervals between successive droppings. After Prop & Vulink (1992).*

represented by food that can be processed at high rates (high IR) and that are easily digestible (high ME), thus enabling the geese to acquire high productive energy. Digestive capacity is the maximal amount of food that can be processed by the digestive tract per unit time. When food intake is constrained by the digestive capacity, this is referred to as a *digestive bottleneck*. The importance of a digestive bottleneck in animal nutrition has been realised for a long time (Adolph 1947), though the relevance for goose foraging dawned much later (Drent & Swierstra 1977, Sedinger & Raveling 1988). If geese ingest food at rates exceeding the

Figure 11.3. *Barnacle Goose ingesting a poor quality food in summer; in this case, large bites are taken from a moss carpet on the Svalbard tundra. Note the bulging oesophagus. This male is foraging during a brief recess from its territory in a nearby breeding colony. Photo by Jouke Prop.*

digestive capacity, it is temporarily stored in the oesophagus, and a foraging break is taken (Prop & Vulink 1992). Digestive capacity is determined by the rate of passage of food through the alimentary tract. Higher rates of passage result in more food that can be processed, thus a higher digestive capacity. High rates of passage through the digestive tract are common in geese, with food retention times in winter as short as two hours. However, these high passage rates come with a cost, as digestion is poor. By slowing down the passage rate, and retaining food for longer, geese digest more of the tough cell wall constituents (Figure 11.2). Slow passage rates yield high energy assimilation during the Arctic summer, when permanent daylight enables geese to distribute foraging time over the full 24h cycle (Prop & Vulink 1992). As an additional benefit of long retention times, geese are able to extract sufficient energy from poor quality plants that are abundant in the high-Arctic environment. Mosses, for example, are widely available and serve as an important source of energy when other plants have not yet emerged from snow cover (Figure 11.3, also Figure 2.3).

Seasonal changes in food

A key period in the year is represented by the spring months when Barnacle Geese are staging in Norway. During this time the geese experienced a massive peak in productive energy (Figure 11.1D), which reflected the pattern of quality and abundance of their main food plants. When the birds arrived, food biomass was low, but halfway through the staging period food abundance increased to maximum values with protein in the food remaining high. In the later part of the spring staging period both food quality and food abundance declined, which resulted in the reduction in productive energy.

Observations made for Barnacle Geese during the pre-migratory period seem representative for geese in general; during each phase of the annual cycle food abundance and quality follow a strikingly similar pattern (e.g. Prop & Deerenberg 1991, Lepage et al. 1998, van der Graaf et al. 2006). The decline in food quality following the spring flush is an almost universal trend in ageing plants (Gill et al. 1989), also experienced during the breeding season in the Arctic (see below) and in temperate areas. The decline in food abundance in the systems we studied was mainly caused by a reduction in the number of suitable leaves due to intensive goose grazing (Prop et al. 1998). It thus appears that the period available to generate productive energy (Figure 11.1) is constrained initially by the low abundance of food and then by a combination of declining food quality and depletion.

As a consequence of seasonal fluctuations in food quality and abundance, high energy requirements are met during sharply demarcated periods of the year. By moving progressively north in spring (Solway Firth – Norwegian coast – Svalbard), Barnacle Geese encounter a spring flush of plant growth three times. This supports the concept of the green wave (Owen 1980a), suggesting that during spring migration geese follow the early flush of growth and by migrating northwards they are able to benefit from successive peaks of digestibility (Drent et al. 1978).

The close association between geese and the phenology of their food further emerges from the spring distribution of the Arctic-breeding goose species across Europe. Each species has its specific breeding range in the circumpolar area, and migration routes differ accordingly. Most strikingly, however, the spring staging sites that are visited just before the geese enter the Arctic climate zone are located within a narrow belt across the continent (Figure 11.4). They arrive within this belt as temperatures reach the lower threshold of grass growth (3–6°C, Vine 1983, Chapman *et al.* 1983), which suggests that during spring migration all goose populations select areas that are similar in the phenological stage of their food. The only apparent exception is the Dark-bellied Brent Goose, which – possibly associated with its original maritime habits – stages in May in a warmer zone than the other species. Brent Geese cope with the inevitable drop in quality of individual food plant species by taking advantage of the successive emergence of saltmarsh plants (Prop & Deerenberg 1991).

The picture above reflects an 'ideal' situation when geese are able to find the most productive location on their way towards the breeding grounds. However, how is this pattern affected when a population keeps growing? This is shown well in the Russian population, which approached a million individuals after a tremendous tenfold increase in 25 years (Chapter 14). Van der Graaf *et al.* (2006) made meticulous measurements of availability and production of food plants at several places along the migration route of this population, and concluded that in the recent past Russian Barnacle Geese benefitted from a green wave, similar to the findings in the Svalbard Barnacle Goose population. This was in agreement with the high deposition rates of body stores by Barnacle Geese staging in the wintering area (Dutch Wadden Sea in March/April) and in the stopover area in the Baltic (in May) as established throughout the 1980s by Ebbinge *et al.* (1991). However, in the early 2000s goose numbers in the Baltic staging (and breeding) area had increased to such an extent that, due to the extremely high grazing pressure, the benefits of the area had been lost (van der Graaf *et al.* 2006). Indeed, from the 1990s onwards an increasing proportion of the

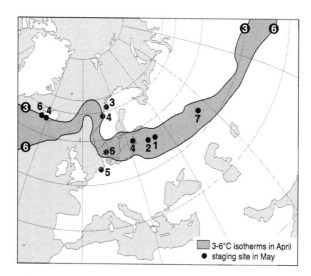

Figure 11.4. *Distribution of goose species across Europe during spring migration (end of April–May) in relation to spring phenology as indexed by the 3–6°C isotherms, the temperature at which grass growth begins. Average temperature data for April are based on 405 weather stations in northern Europe (extracted from Müller 1980). Geographical centres of distribution of high-Arctic populations (based on Madsen et al. 1999) are indicated: 1 = White-fronted Goose, 2 = Tundra Bean Goose* Anser fabalis rossicus, *3 = Pink-footed Goose, 4 = Barnacle Goose, 5 = Dark-bellied Brent Goose, 6 = Light-bellied Brent Goose, 7 = Red-breasted Goose. From Prop (2004).*

Russian population skipped the Baltic as a staging area. Instead, more individuals stayed in the Wadden Sea area for a longer time to prepare for migration (Eichhorn *et al.* 2009). Rather than a two-step migration towards the Russian sub-Arctic the geese travelled the distance without substantial stopovers (Eichhorn *et al.* 2009).

Optimal timing of breeding

Reproduction of Arctic-breeding geese is timed to provide goslings with the most nutritious and abundant food (Owen 1980a). We tested this idea by conducting an analysis similar to that described for the spring staging area (above), but this time we used food quality and abundance data from the breeding grounds in Svalbard. Similar to the pre-migration situation, the large increase and subsequent decline in summertime productive energy (as seen in Figure 11.5) was the result of two opposing tendencies with regard to goose food. Initially the rapid increase in available food outweighed any decrease in food quality, followed by a decline in both food abundance – because of depletion – and food quality. In contrast to the pre-migration period, the peak in productive energy was relatively early, and was already at its maximum on day 12 of the 42-day brood-rearing period. This means that newly hatched goslings experienced prime foraging conditions during the first two weeks after hatch; after that, conditions quickly deteriorated with the period of greatest energy requirements for goslings still to come (at 11–25 days of age, Lepage *et al.* 1998). It seems, therefore, that average hatch dates were later than expected on the basis of foraging conditions during brood rearing alone.

The early peak in productive energy seems to have extremely important fitness consequences, as the number of young raised through autumn strongly declined with later hatching (Figure 11.6B). The trend was almost entirely determined by gosling survival because the number of eggs hatched did not vary with laying dates; the larger size of early

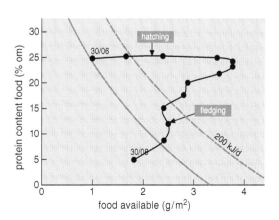

Figure 11.5. *Seasonal changes in food resources (the grasses* Dupontia fisheri *and* Festuca rubra *combined) during brood rearing in Svalbard, represented by quality (y-axis) and abundance (x-axis) for successive five-day periods. Average dates of hatching and fledging are indicated. The solid line indicates the combinations of food quality and abundance when geese are in energy balance. The dashed isoline indicates when surplus food intake yields a productive energy of 200kJ/day.[93] From Prop (2004).*

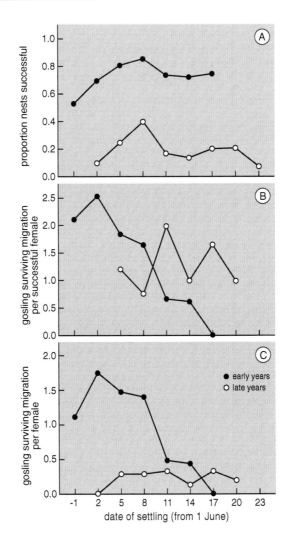

Figure 11.6. *Reproductive success in relation to the date of settling (onset of egg-laying) in the Svalbard population of Barnacle Geese (recalculated from Prop & de Vries 1993). (A) Probability of hatching at least one egg. (B) Number of offspring surviving autumn migration per successful female. (C) Number of offspring per female initiating a nest. Data are for early (n = 251) and late (n = 258) snowmelt years.*

clutches (Dalhaug *et al.* 1996) was counterbalanced by more egg losses in early nests. Potentially, the trend in young produced could have been affected by a decline in parent quality as well, but evidence for this was weak (see next section, also Lepage *et al.* 1999).

The probability of hatching the eggs increased with later dates of settling in the colony (Figure 11.6A). This means that fitness benefits from early breeding were offset by a low nest success, at least for the very early individuals. Apparently nest initiation was a trade-off between opposing benefits for the clutch and goslings (Prop & de Vries 1993). On balance, however, evidence suggests that there was strong selection for breeding early (Figure 11.6C), yet – given the early drop in food quality during brood rearing – average breeding dates were relatively late. This seemingly counter-intuitive pattern is also found in other Barnacle

Goose populations (van der Jeugd *et al.* 2009), in other goose species (Lepage *et al.* 2000) and in other avian taxa (Daan *et al.* 1990). There are three explanations as to why individuals would breed at dates that at first glance appear suboptimal.

First, egg-laying dates may be constrained by limited availability of nesting sites (Perrins 1970, Sedinger & Raveling 1988, Gauthier *et al.* 2003). Although this may play a role in species breeding in habitats covered by spring melt floods, it is unlikely that this explains late breeding in Barnacle Geese because nests are often on windswept places not covered by water or snow.

Second, there may be a fitness cost associated with early breeding. In some years, early arrival on the breeding grounds may have consequences for an adult's future survival. In one extremely late spring, Barnacle Geese in Svalbard waited 30 days for the snow-covered breeding grounds to thaw before initiating egg-laying. It seems plausible that in such years survival probabilities for individuals attempting to breed early were depressed, though this could not be measured.

Third, date of egg-laying might be constrained by limited food resources encountered en route to the breeding grounds (Sandberg & Moore 1996). This is supported by evidence from the long-term study of Greater Snow Geese (Bêty *et al.* 2003, Dickey *et al.* 2008). In years with adverse weather conditions these birds deposited less body stores and started nesting later than in normal years. The situation would be compounded in years when the window of time available for depositing pre-migratory body stores is late relative to conditions on the breeding grounds (a 'mismatch' in timing; Dickey *et al.* 2008, Saino *et al.* 2011, Clausen & Clausen 2013). Barnacle Geese breeding in Svalbard seem to be headed towards a mismatch in timing as spring temperatures are increasing in the wintering area (UK), the stopover areas (Norway) and in the breeding area, whereas no change in timing of migration has been detected (Tombre *et al.* 2008).

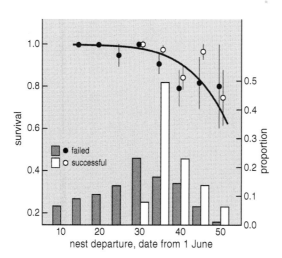

Figure 11.7. *Seasonal survival probability (June–December) in relation to the date of nest departure for failed and successful breeders at the Diabasøya colony on Nordenskiöldkysten.[94] For comparison, the frequency distributions of the date of hatch (successful individuals) and dates of abandoning the nest (failed individuals) are given (right y-axis). Plots indicate a lower survival probability for individuals departing the nest later in the season. Averages ± SE are indicated. Details in Prop* et al. *(2004).*

There is ample evidence from other goose study systems that late hatching incurs a fitness cost to both parents and offspring (Sedinger & Raveling 1986, Manseau & Gauthier 1993, Lepage *et al.* 1998). Late-hatched goslings have a lower growth rate than those hatched earlier (Sedinger & Flint 1991, Lindholm *et al.* 1994), which results in lower survival (Williams *et al.* 1993, Lepage *et al.* 1999, van der Jeugd *et al.* 2009). Similarly, adults starting the wing moult late have a low probability of surviving. Although nutritional stress can be expected to be less in adult geese than in goslings, adults' survival rate during the year after the breeding episode is inversely correlated with the date of hatch (Figure 11.7). A late onset of the breeding season is heavily penalised; survival of females with late-hatching clutches (0.80) was dramatically lower than for those that bred earlier (0.98). Once geese have decided on a particular date for egg-laying this sets the schedule for the rest of the summer, and hence the probability of survival. Only abandoning the clutch, and proceeding with the moult early in

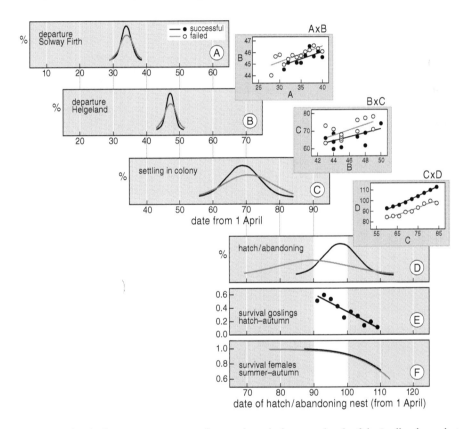

Figure 11.8. *Individual variation in timing of events through the annual cycle of the Svalbard population of Barnacle Geese. Given are frequency distributions (%) for individuals producing and hatching a clutch (black lines) and those that fail to hatch eggs (grey lines). Events are: (A) departure from Solway Firth in the UK; (B) departure from the staging areas in Norway; (C) settling in the breeding colony in Svalbard; (D) hatching the clutch (successful breeders) or abandoning the nest (failed breeders). Inserts show correlations between timing of successive events for successful and failed breeders.[95] (E) Survival of goslings from hatch through autumn migration in relation to date of hatch. (F) Survival of females through autumn migration in relation to date of hatch or abandoning the nest. The window of time in which eggs should hatch to ensure maximal survival of goslings and adults (dates 90–100) is indicated. From Prop (2004).*

the season, ensures sufficient recuperation of body condition and nutrient stores to avoid a reduction in survival during autumn migration (Prop *et al.* 2004).

To gain insight into the window of opportunity in which Barnacle Geese can lay their eggs with reasonable fitness prospects, data on survival of goslings and adult females are summarised in the two lower panels of Figure 11.8. The declining survival potential of goslings and females effectively reduces the seasonal laying window to a mere 10 days (indicated by the shaded bar in Figure 11.8), which equates to only half of the three-week laying period actually used by individual geese. Therefore, the timing of nest initiation seems all-important.

Individual strategies within the annual cycle

What might cause variation in timing of breeding among individuals, and why are some individuals apparently breeding at a time that impairs their fitness? One possible reason is that individual performance depends on what has been achieved earlier in time (Piersma & Baker 2000, Newton 2004); producing a large and timely clutch requires an early preparation in spring. These 'carry-over' effects might play an important role in migratory birds in particular, as they are usually constrained by food through much of the year. To test if individuals were consistently early or late in the annual cycle, we compiled observations of marked Barnacle Geese on key sites along the flyway. The first of the three small panels on the right hand side of Figure 11.8 shows the relationship for individuals' departure dates from the Solway Firth and the Norwegian coast (panel marked A×B). The second (B×C) and third (C×D) panels show the relationships between departure from Norway and settling in the breeding colony, and settling in the colony and the date of either hatching or abandoning eggs. All three relationships are significantly positive, which means that individuals are consistently early or late in each stage. This is congruent with the observation that early breeders in Svalbard were traced back as belonging to the earliest migrants from the Solway Firth (Prop *et al.* 2003). It seems, therefore, that individuals have a limited opportunity to 'correct' for poor timing, and breeding at the right time has its basis in the wintering grounds, several months before the first egg is laid (Sedinger *et al.* 2011).

The individual's decision on timing of breeding is thought to result from a trade-off between the benefits of breeding early and of waiting in order to gain more stores to invest in additional eggs (i.e. the *individual optimisation hypothesis*; Drent & Daan 1980). There is no reason to think such a trade-off is unique for the breeding season because decisions made earlier in the year affect fitness as well. Individuals can be expected, therefore, to trade off the prospects of depositing additional body stores against the benefits of early migration or reproduction, and choose the timing of migration accordingly ('choice' in the left part of Figure 11.9). Successful foragers might be expected to postpone migration because they have the best opportunities to gain additional stores, and they should not arrive too early on the breeding grounds. In contrast, poor foragers should depart before they have maximal stores because they have poor prospects for gaining additional stores and they should not arrive too late.

Observations in the Norwegian spring staging area supported the idea that timing of migration is tuned to the individuals' foraging abilities; poor quality individuals, which

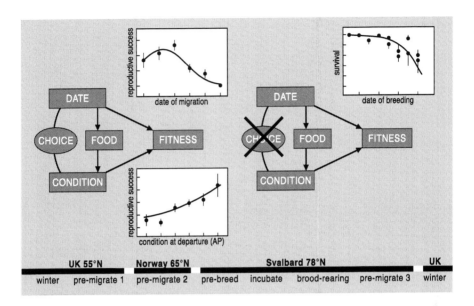

Figure 11.9. *Illustration showing the individuals' trade-off between timing of events in the annual cycle and body condition. For spring (left part of figure) this is described by relationships between reproductive success and date of migration (upper graph) and body condition at departure (lower graph, Prop et al. 2003). For summer this is indicated by the relationship between adult female survival and date of breeding (right part of figure). See text for explanation. From Prop (2004).*

achieved a low intake rate of food, migrated earlier than better performing birds (Prop *et al.* 2003). Remarkably, we found no evidence for a similar quality gradient in nest initiation dates. Rather than initiating a nest early, poor-performing individuals spent more time foraging near the breeding colony before settling (Prop *et al.* 2003, Hübner *et al.* 2010). Early migrants might benefit from a prolonged pre-breeding period in Svalbard, as the low densities of foraging geese at this time of the year might make them less susceptible to interference and depletion of food than during the migration period (Fox & Bergersen 2005). Other individuals appeared to spend almost no time on pre-breeding sites and arrived on the nest site almost without any delay after arrival in Svalbard (Griffin 2008). It seems, therefore, that similar trade-offs may result in different outcomes depending on the time of the year.

Summer ends abruptly when a rapidly decreasing day length in combination with autumn snow forces the geese to migrate south. This inflexible ending adds to a rigid time schedule with little variation in the period required for incubation and brood rearing; the entire period takes 72 days from egg-laying to fledging in most Arctic-breeding geese (Owen 1980a). The opportunity early in the year to adjust the timing of migration and reproduction to body condition thus switches to a stranglehold of a preset time schedule once the clutch has been produced (i.e. 'no choice', right part of Figure 11.9). Individuals initiating a late nest cannot avoid poor foraging conditions in the remaining summer and a short pre-migration period; as a consequence, body stores in autumn are bound to be inadequate. Thus, poor timing in June during the early breeding season is heavily penalised in September when survival during the autumn migration depends on pre-migratory fat and nutrient stores.

Summary and conclusions

This chapter started with investigating the effects of abiotic (day length) and biotic (food quality, food abundance) factors on energy intake. To achieve a positive energy balance all factors must be complementary, especially during periods when geese need most energy: in spring when preparing for migration and reproduction, in midsummer during the development of goslings, and in late summer before the southward migration. Geese extend the period of favourable foraging conditions in spring by moving northwards and benefiting from successive flushes of spring plant growth (the *green wave hypothesis*). The period available to generate productive energy is constrained initially by the low abundance of food and then by a combination of declining food quality and depletion. Continuous daylight in the Arctic enables geese to retain food for longer periods in the digestive tract, thus enhancing the assimilation of nutrients. In summer geese can afford to ingest food that is of lower quality than in winter, and the longer time available for foraging still enables them to achieve a high productive energy.

The timing of reproduction is a trade-off between hatching eggs successfully (by choosing an incubation period with favourable conditions) and survival of the goslings (such that brood rearing coincides with the best period for gosling growth). Producing a late clutch has a considerable cost to females because of a decreased probability of survival in the following year. The window of time to initiate a nest on the Arctic breeding grounds, with reasonable prospects of producing surviving goslings through the autumn without affecting the survival probability of breeding females, is a mere 10 days. Within this short time frame early breeders are usually most successful. To manage an early start of breeding, the migration must be scheduled accordingly. An analysis using resightings of marked individuals from key sites in the flyway revealed that individuals were consistently early or late in each stage of the annual cycle, starting with migration from the wintering grounds to the date of nest initiation. Once settling in the breeding colony and initiating a nest, the time schedule is set for the rest of the summer, with relatively fixed periods needed for egg-laying, incubation, brood rearing and moult (totalling 2.5 months). The lack of opportunities to catch up in the short Arctic summer results in a strong selection for early breeding.

Statistical analyses

[93] Energy balance was calculated using the following parameters and equations: daily energy expenditure (DEE) is 1,000kJ/day (Ebbinge *et al.* 1975, Drent *et al.* 1978); intake rate (g/min) = 0.074 + 0.0041 × food available (g/m^2) (Chapter 8); feeding time is on average 900 min/day (Prop & Black 1998); protein content (% of food) needed to generate a particular metabolisable energy (ME) is –15.04 + ME × 5 (Figure 11.5) (Prop & Vulink 1992, Prop *et al.* 2004).

[94] Seasonal survival probability in relation to the date of nest departure for failed and successful breeders: y = 11.67 – 0.156x (on a logit scale); χ^2 = 12.22, df = 1, P < 0.0001; based on a logistic regression. See Figure 11.7.

[95] Correlation coefficients between timing of successive events for successful and failed breeders, respectively: A×B r_{74} = 0.38 and r_{279} = 0.38; B×C r_{12} = 0.52 and r_{10} = 0.63; C×D r_{231} = 0.95 and r_{208} = 0.53; P < 0.05 in all cases. See Figure 11.8.

CHAPTER 12

Site fidelity and movements

When Barnacle Geese were discovered in Svalbard their nests were found only on cliff faces and rocky slopes in the mountains (Jourdain 1922, Løvenskiold 1964), sites that were safe from Arctic Foxes. Barnacle Geese are renowned for their ability to nest on precarious rock ledges and pinnacles from which goslings make death-defying jumps to the sea or rock scree below (Cabot 1984, Mitchell *et al.* 1998). In the 1950s the birds began to make use of offshore, fox-free islands (Prestrud *et al.* 1989). By 1960 there were 11 colonies, increasing to 32 by 1980, and the number has grown considerably since then (Figure 1.7). The number of goose nests at the colonies themselves has also continued to increase, even on the small, rocky islands that already seemed quite full. For example, in a 20-year period we observed a fourfold increase in the number of nests in the Nordenskiöldkysten colonies, making it extra challenging for late-arriving pairs to squeeze in between already established pairs (Chapter 7). It became obvious that while many geese chose to return to the site where they were born (natal philopatry) and compete with the home crowd for space to nest, others chose a more uncertain future and dispersed to new locations to establish a breeding career.

The dilemma that geese face regarding the choice to return to the site where they were raised or where they bred the previous year should depend on breeding opportunities. Horn (1984, p. 61) argued that 'Different degrees of competition and crowding at various stages of life are important in determining the likelihood of an individual establishing itself locally versus at a distance. This in turn determines the adaptive value of dispersal.' Therefore, increasing population density will probably affect the decision to return or not. Pioneering geese that discover new sites where there is less competition for resources may be able to surmount the costs of dispersal. Costs of dispersal might involve extra travelling, lost time, costs related to settling, food finding, avoiding a new set of predators, and breeding in an unfamiliar environment among unfamiliar individuals.

The consequences of dispersal and philopatry might differ between the sexes, causing a disparity in male and female dispersal rates. For example, natal dispersal is nearly always female-biased in birds, although the opposite is observed in waterfowl. Greenwood (1980) suggested that this is because the mating system in waterfowl is based on mate defence rather than resource defence, in contrast to most other bird species. The benefits of experience or 'knowledge' of the breeding area might be greater in female than in male waterfowl because of their larger energetic investment during the breeding season (Lessells 1985, Rohwer & Anderson 1988).

In this chapter we describe the characteristics and fate of geese that return to the natal or breeding area and those that move to new breeding locations. Methods that were involved are described in Box 12.1.

BOX 12.1 DISPERSAL AND PHILOPATRY METHODS

During the summer a dispersal event is said to occur when a bird's breeding location is different from the previous year. Philopatry is when an individual remains in the same location between years. Workers using the term natal dispersal or natal philopatry provide information with regard to a bird's natal (birth) site and its first known breeding area, and breeding dispersal refers to the location of subsequent breeding sites.

We measured dispersal and philopatry in two ways: the *local return* and *dispersal destination* methods. Both methods are challenging and require much effort in resighting marked individuals, and both have notable shortcomings (Lebreton *et al.* 2003).

The local return method, employed in the Baltic and Svalbard studies, requires observers to 'take attendance' within a breeding area for several years to determine the number that do and do not return. A problem with this method is that the fate of missing birds is unknown. One way to solve this problem is to devalue return rates by the number of birds that are still alive, which is possible by checking subsequent resighting records made on the wintering grounds. In this way only those individuals known to be alive are included in the analysis. It is possible, therefore, to calculate dispersal from breeding areas for surviving birds by checking for those that did not return.

The dispersal destination method was used in Svalbard where birds were resighted at their 'destination' in a subsequent summer in either the original or new breeding area. Between 1973 and 1995, field teams in Svalbard amassed 15,729 observations of 8,135 individuals during 16 capture expeditions and five multi-colony surveys among 19 breeding areas (Table 12.1). From this effort, we managed to relocate 2,207 birds in a second year. An advantage to using this method is that dispersal distances can be measured. A problem with this method is that the precise year of the movement is unknown and many birds go undetected since it is difficult to spend enough time at all potential dispersal destinations.

	Latitude °N	First record	Years with recoveries	Total birds recovered
Dunøyane[a]	77°03'	1938	11	560
Tusenøyane[b]	78°09'	1987	3	7
Isøyane[c]	77°09'	1938	5	366
Eholmen[d]	77°36'	1987	1	14
Mariaholmen/Akseløya[e]	77°41'	1989	2	14
Reiniusøyane[f]*	77°45'	1965	10	1,136
Diabasøya[g]*	77°46'	1968	13	1,944
St Hansholmane[h]*	77°51'	1964	8	1,808
Reindalen[i]	77°52'	1954	1	3
Daudmannsodden[j]	78°13'	1970	4	346
Erdmannflya[k]	78°18'	1980	1	16
Forlandøyane/Prins Karls Forland[l]	78°20'	1963	3	87
Sassendalen[m]	78°21'	1963	5	117
Bohemanflya[n]	78°23'	1980	1	160
Gåsøyane/Gipsdalen[o]	78°27'	1982	3	407
Hermansøya[p]	78°33'	1989	1	1
Kongsfjorden/Ny-Ålesund[q]	78°56'	1980	10	1,130
Kapp Mitra[r]	79°07'	1984	1	2
Moseøya/Danskøya[s]	79°39'	1984	1	17

* These colony areas are located on Nordenskiöldkysten.

Table 12.1. *Place names listed from south to north where 8,135 individually marked Barnacle Geese were observed in Svalbard. Recoveries refer to birds that were captured, seen or recovered dead. First record refers to the first year a Barnacle Goose subpopulation was discovered at the site. Totals refer to individual birds that were recovered, indicating relative study effort in different areas prior to 1998. After Black (1998a). Refer to map in Figure 12.2 for locations indicated by superscript letters.*

Philopatry and dispersal in the Baltic

We employed the local return method for 10 cohorts in the main Baltic study colony on Gotland. This colony was intensively monitored each summer between 1984 and 1998 when the number of nests increased from 325 to 2,220. Figure 12.1 shows the proportion of surviving geese that did not return to the natal colony to breed once they were old enough to do so. Young males had a higher dispersal rate than young females. Male dispersal rate increased from around 45% in the beginning of the study to around 65% at the end, while female natal dispersal rate remained low at about 20%. The increase in male natal dispersal rate could most likely be explained by increased competition for space and mates and by increasing numbers of potential partners from other Baltic colonies (van der Jeugd &

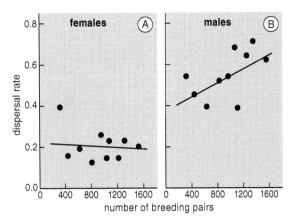

Figure 12.1. *Natal dispersal rate for (A) females and (B) males in relation to the number of breeding pairs in the main Baltic study colony on Gotland. Natal dispersal rate of males increased as number of breeding pairs increased. Dispersal rate of females was consistently lower than for males. From van der Jeugd & Larsson (1999).*

Larsson 1999). However, because natal dispersal rate, in absolute terms, was already higher for males than for females at the beginning of the study when population density was still low, the difference between the sexes must also be due to other reasons (see Lindberg *et al.* 1998 for a similar trend in Black Brant in Alaska). Remarkably, Barnacle Geese that were philopatric were frequently observed in the 'home' colony in the 'yearling' summer. This initial exploration of the natal colony (referred to as prospecting) may be important for gathering information about competition and distribution of local resources.

That females did not disperse more with increasing densities supports the notion that they value the benefits of philopatry more than males. Information about the location of good foraging sites in the breeding area may be advantageous, particularly to females during incubation breaks when they strive to replenish depleted body reserves (Prop & de Vries 1993). Females, more than males, may also garner non-trivial benefits from kin or familiar neighbours (Chapter 7), which may not be as available to those that disperse.

Philopatry and dispersal in Svalbard

Once geese found their long-term mates and began a breeding career at a particular colony, they rarely dispersed, even when unsuccessful. Based on measures of local return rates for two Svalbard colonies, dispersal away from breeding colonies for adults was calculated at less than 2% in the 1980s on Nordenskiöldkysten (Prop *et al.* 1984) and about 12.5% in the 1990s at Kongsfjorden (Loonen *et al.* 1998). Using the dispersal destination method, we found that fewer adults were relocated outside their original capture site than young birds; only 87 (6%) of 1,510 individuals ringed as adults, compared to 115 (16%) of 697 ringed as goslings were found elsewhere,[96] suggesting a stronger movement tendency in younger birds. Based on the sample marked in their gosling year, males changed sites more than females (26% of 288 males, and 10% of 409 females).[97] For the most part, therefore, philopatry was the favoured strategy in the Svalbard population, though dispersal gained popularity, especially for younger birds and more so for males than females. By comparison, in the long-term study of Black Brant at the Tutakoke River colony on the Yukon–Kuskokwim Delta, Alaska, 17% of females did not return to the colony prior to breeding age, but for those that did return and establish their first nests, subsequent return rate was 100% (Sedinger *et al.* 2008).

Dispersal distance

Figure 12.2 shows the dispersal destinations for geese departing from the Nordenskiöldkysten study area. They dispersed variable distances (25–225km) to sites in all directions. Birds marked as goslings moved larger distances (average 98.2km, SE 8.7, n = 41) than those marked as adults (average 69.8km, SE 6.4, n = 28).[98] Twenty-four marked birds (seven females and 17 males) were recorded in new colonies, suggesting that they were colony founders (Black 1998a). Dispersal distances were not significantly different between the sexes.[99]

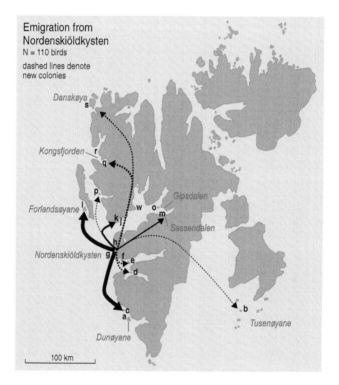

Figure 12.2. *Destinations of 110 dispersal events determined by recovery/resightings of marked geese outside their initial area, at Nordenskiöldkysten in Svalbard. between 1977 and 1996. Arrow and line size reflect relative number of movements to distant sites (range 1–25 geese). Dashed lines indicate movements to newly established colonies. Groups of up to 16 marked geese from Nordenskiöldkysten were relocated at new destinations; 44% of movements were in groups of five or more marked individuals. Two females temporarily moved from this natal site, returning after one or two years. Letters refer to the areas mentioned in Table 12.1. After Black (1998a).*

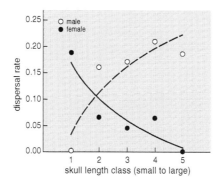

Figure 12.3. *Dispersal rate in relation to structural body size (indexed by head length) of female and male Barnacle Geese from Nordenskiöldkysten in Svalbard. Dispersal was more likely in larger males and smaller females. Skull length sample sizes ranged from 37 to 60 for each female point (total n = 223), and 19 to 37 for each male point (total n = 154). Categories ranged from small heads (81.7mm females, 85.8mm males) to large heads (85.9mm females, 90.0mm males).*

Natal dispersal and body size

Access to different amounts and types of food resources during the first weeks of life influences gosling growth and final adult body size in geese (Chapter 10). In all groups that we managed to capture, measurements always showed a wide range of body sizes. We explored the Svalbard data to test whether body size was linked to the probability of moving to a new area. Figure 12.3 indicates that the probability of moving to a new area increased with male body size while it decreased with female body size.[100]

Why would some of the larger males and smaller females choose to move to a new site as opposed to staying in a familiar site where kin and other associates live? The majority of individuals returning to large or growing colonies will presumably experience increasing difficulties becoming established due to competition for nests and food. Young birds that manage to establish nests at natal sites often do so on the outer edges of the colonies where predation pressure is likely to be higher and nest success lower (Finney & Cooke 1978). Large males that return to their natal sites in their first summers may have an advantage in social encounters, but the pay-offs for winning there may not be great. Henk van der Jeugd (2001), who studied this problem in the Gotland colonies, argued that larger males may have more success at establishing a territory at a new site than when returning to a crowded natal colony. Furthermore, smaller males that return to natal sites may benefit from familiarity with habitats and neighbours, which may allow them to become established in the colony sooner than if they moved to a new location.

Another possibility influencing dispersal tendencies in different sized birds may be related to pairing chronology. In Chapter 4 we provided evidence to suggest that large females were preferred as mates; they participated in more trial liaisons and paired earlier than smaller females (Choudhury & Black 1993). Larger females may, therefore, remove the best quality males from the available pool of mates, leaving smaller females with a suboptimal choice of mates if they stay at the natal site. Small females, therefore, may improve on their choice of potential long-term partners by moving to a new site, a benefit that may, in the long run, compensate for costs of dispersal.

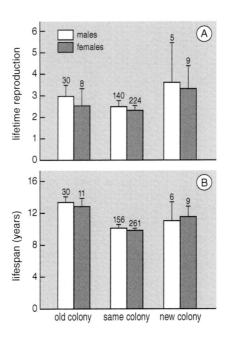

Figure 12.4. *Comparison between dispersal types from an early established, high-density colony (Nordenskiöldkysten) of (A) lifetime reproductive success (number of goslings returning to wintering grounds) and (B) lifespan (years). Dispersal included movements to 'old' colonies, returning to the same area, and pioneering newly established sites. Averages, SE and number of birds are indicated.*

Consequences of dispersal

If all else was equal, geese that settle in favourable sites should outperform those that settle in unfavourable sites. However, we have shown in earlier chapters that there are many inequalities among breeding areas. We described how colonies vary in vegetation characteristics, microclimate, predation risk and density of geese (Chapter 9, 10, 14). Whereas older colonies may be at or close to saturation, newer colonies are less constrained by density-dependent processes, offering immigrants a better chance of breeding successfully. On the other hand, individuals that do move to new areas probably face substantial costs that limit reproductive prospects. We compared the fate of Nordenskiöldkysten geese in relation to dispersal strategy, including movements to (i) another well-established colony (general dispersal), (ii) back to the original colony (philopatry), or (iii) a new colony (colonisation).

Figure 12.4 shows the average lifetime reproduction and lifespan for birds exhibiting each of these dispersal options. The comparison shows that lifetime reproduction was slightly higher for birds that moved to new colonies; however, the power of this test was insufficient to detect clear differences in the three dispersal types.[101] Instead, variation in lifespan was significantly different among dispersal types for males,[102] with the longest lives for birds moving to other sites and the shortest lives for philopatrics (Figure 12.4B). This suggests that individuals that dispersed performed at least equally well as those that stayed at natal sites.

Mechanisms of dispersal

Not much is known about the actual mechanisms that trigger long-distance dispersal and explorative behaviour in waterfowl. However, we suspect it is much the same as mechanisms influencing movement among smaller food patches. A likely day-to-day cue involved in the decision to stay or leave is foraging success. Geese may use the simple rule of stay if food finding exceeds a threshold of acceptance. If foraging performance falls below the acceptable threshold, for example at a spring staging location, the goose should either leave immediately or it should not return the next year. In Chapter 8 we described the benefits of returning to a 'known' site in terms of acquiring food. On return to sites geese reoccupy rich food patches that were discovered in previous years and enjoy enhanced access to food that comes with an increment in social status due to the prior residence effects. Few geese would return to sites where feeding bouts are interrupted by travel time spent searching for food. In the same way that prolonged searching for food would reduce intake rates, foraging success would also be negatively affected by too many disturbances to foraging bouts by aggressive flock members, persistent predators or human disturbance, encouraging dispersal to new sites. In addition to finding food, other cues leading to the tendency to disperse in young geese may be the frequency of encountering trial partners, or nest establishment opportunities. Predation risk during the nesting phase is also likely to shape dispersal tendencies (Spaans *et al.* 1998, Quinn *et al.* 2003).

Day-to-day experiences like these may be linked to the production of physiological responses that drive tendencies for pioneering into unexplored landscapes. In other animals dispersal probabilities have been linked to numerous environmental factors affecting habitats (biotic and abiotic) and to internal conditions of individuals (i.e. measurable behavioural, morphological and physiological traits, reviewed by Ims & Hjermann 2001). For other birds there is evidence that natal dispersal is influenced by juvenile body size/condition or by food in the parental environment (Marsh Tits *Parus palustris*, Nilsson 1989; Spanish Imperial Eagles *Aquila heliaca,* Ferrer 1993; Western Screech Owls *Otus kennicottii*, Belthoff & Dufty 1998). It has also been suggested that female modification of steroids deposited in their chicks' egg yolk might provide the physiological basis for offspring behavioural profiles (aggressiveness, activity, social tendency; Cote *et al.* 2010), thus shaping natal dispersal from overcrowded or resource-poor environments (Dufty & Belthoff 2001, Duckworth 2009). Research designed to uncover the link between physiological responses and dispersal/homing behaviour in waterfowl would help us better understand the processes involved in site fidelity, dispersal, colonisation of new sites and exchange among populations. In the next chapter we describe the occurrence and consequences of immigration and emigration between adjacent Barnacle Goose populations.

Summary and conclusions

Costs of dispersal might involve those related to settling, food finding, predation risk, and breeding in an unfamiliar environment among unfamiliar individuals. Breeding opportunities are expected to influence philopatry and dispersal tendencies. Knowledge of breeding sites and benefits gained from sharing with kin should promote high degrees of philopatry. However, benefits of returning to natal sites will be reduced by increased competition and crowding in increasing populations (as in the Barnacle Goose populations). In Chapter 14 we provide evidence of reduced reproductive success as colonies grew larger and that smaller, recently established colonies actually produced more goslings per breeding pair. In the current chapter we showed that while the majority remained philopatric, young geese dispersed to alternative breeding sites more often and at greater distances than older geese. Male natal dispersal from the main Baltic study colony increased from around 45% in the beginning of the study to around 65% at the end, while female natal dispersal rate remained low at about 20%. Dispersal distances ranged from 15 to 225km (average 93km) in the Svalbard archipelago. Evidence suggests that large males and small females were more likely to disperse. Unlike larger males, small males may not be as able to surmount the extra costs of dispersal. We speculate that small females that disperse from natal sites have a better chance of finding suitable long-term mates than those that remain at overcrowded natal areas. There was little difference in lifetime reproduction between geese that remained at natal sites or dispersed to other colonies. However, males that moved to other breeding areas enjoyed a longer lifespan, demonstrating a benefit to dispersal from crowded sites. We assume that geese make decisions about returning to sites over days within seasons, and over longer periods of time from one season to the next. The actual mechanisms that drive dispersal in geese are still largely unknown.

Statistical analyses

[96] Frequency of dispersal and philopatry between those captured as adults and goslings ($\chi^2 = 66.1$, df = 1, $P < 0.0001$). Based on dispersal destination method.

[97] Frequency of dispersal and philopatry between males and females ($\chi^2 = 30.16$, df = 1, $P < 0.001$). Based on dispersal destination method.

[98] Dispersal distance comparison between younger and older geese (t = 2.40, df = 67, $P < 0.02$). Based on dispersal destination method for males from the Nordenskiöldkysten breeding area.

[99] Natal dispersal distance comparison from Nordenskiöldkysten between the sexes (males: average 86.7km, SE 6.0, n = 69; females: average 76.2km, SE 7.1, n = 41; t = –1.12, df = 108, NS). Based on dispersal destination method.

[100] Logistic regression of the probability of dispersal by body size (males n = 154, $\chi^2 = 3.86$, df = 1, $P < 0.05$; females n = 223, $\chi^2 = 5.30$, df = 1, $P < 0.05$). This analysis was based on the dispersal destination method. See Figure 12.3.

[101] Lifetime reproductive success was not a function of dispersal type (GLM males $F_{2,173} = 0.53$, NS; females $F_{2,237} = 0.35$, NS). The model included cohort effect (1975–1989), which significantly affected lifetime reproduction. See Figure 12.4A.

[102] Lifespan was a function of dispersal type for males, but not females (GLM males $F_{2,190} = 3.42$, $P = 0.035$; females $F_{2,279} = 1.95$, NS). The model included cohort effect (1975–1989), which significantly affected lifespan. See Figure 12.4B.

CHAPTER 13

Exchange among populations

Although wild geese often follow distinct migration routes, exchange among populations does occur. On one occasion a group of foreign Barnacle Geese landed among a flock of captive Hawaiian Geese and other exotic waterfowl at WWT Slimbridge. The new arrivals consisted of a ringed female with an unringed male and three juveniles. Right away several observers agreed that the slim-looking Barnacle Geese must not belong to the collection as they were extra vigilant and apparently unsure what to do next. We grabbed some binoculars and read the female's engraved ring seconds before the geese took off. Records confirmed that the ring was fitted in Ireland, indicating that the family was several hundred kilometres south-east of its normal wintering range. In this chapter we examine the evidence regarding exchange of individuals among the Barnacle Goose populations.

The migratory range for three of the five populations is impressive: 3,700km from Russia to the Netherlands; 3,100km from Svalbard to the wintering grounds in the UK; and 2,800km from Greenland to Scotland and Ireland (Figure 2.1). The distances between wintering areas are only 180km (Solway Firth to Islay, Scotland) and 600km (Solway Firth to the Netherlands), so exchange between them is feasible. Birds can easily move such distances during exceptional weather conditions (storms or widespread snow cover, Newton 1998). In this chapter we treat the Russian, Baltic and the recently developed North Sea populations as one complex, as they share a common wintering area.

Conservation and management is usually at the level of populations or flyways (Delany & Scott 2006, Kirby *et al.* 2008). Measures of immigration and emigration are used as a guide to the amount of genetic material that is shared between populations, and enables us to consider the likelihood of genetic divergence, and ultimately speciation. To appreciate the long-term consequences of geographical separation of populations, one has to therefore consider the exchange of genetic material (Hewitt & Butlin 1997).

Evidence for movements among populations

The first foreign-marked bird that was detected in the Svalbard Barnacle Goose population was on the wintering grounds in October 1976 during a rocket-net catch. The metal-ringed female had been captured and ringed twice previously, first during an expedition to Jameson Land, Greenland in July 1963 and again in Inishkeas, Ireland in April 1976. For two months after the catch on the Solway, observers at WWT Caerlaverock saw this bird, with its new engraved alpha-coded ring, 15 times together with a mate that had also been fitted with its ring during the rocket-net catch. The pair's stay was apparently short-lived as observers found the birds back in their original range the next winter. This initial record was followed by regular sightings of other birds apparently getting 'blown' off course, particularly during periods of harsh weather. The most exceptional records of Barnacle Geese outside their normal range were recovered in France, Spain and Canada (Ogilvie & Owen 1984, BTO 2013).

The likelihood of detecting foreign birds increased when observers from other European countries began reading engraved plastic leg-rings with spotting scopes. Use of individually coded rings for Barnacle Geese was initiated in 1973 in the Svalbard population, in 1979 in the Russian population, and in 1984 in the Greenland and Baltic populations, and more than 9,000, 2,800, 3,250 and 2,400 alphanumeric plastic rings were fitted, respectively (Ebbinge *et al.* 1991, Ganter *et al.* 1999, Larsson *et al.* 1998, Ogilvie *et al.* 1999, S. Percival pers. comm.). Since so much effort was made to resight marked birds in all populations, it became possible to detect the presence of unusual rings.

A common method for estimating the number of foreign birds in a population is to extrapolate from observations of foreign-marked individuals. Ebbinge (1985) used this method to estimate how many Svalbard Barnacle Geese joined the Russian population wintering in the Netherlands and Germany between 1980 and 1983. The estimate was based on eight records of Svalbard-ringed geese that were resighted in the Russian population. At that time 1,600 of the 8,500 Svalbard geese were carrying engraved rings, which indicates that 0.5% (= 8 / 1,600) had moved between populations. A crude estimate of the total number of geese assumed to have travelled in the same direction as the eight marked birds is 43 birds (= 0.005 × 8,500). We improve on this method below by employing a Markovian transition model, which takes the probability of resighting rings into account (Brownie *et al.* 1993).

Rates of movement among populations

Only 15 foreign-marked individuals were detected in the wintering range of the Svalbard population, which is a small number given the 400,000 sightings of rings during the study (Table 13.1). On the other hand, no fewer than 68 Svalbard-ringed geese were detected in the wintering areas of other populations (Figure 13.1, Table 13.2).

One method to estimate rates of movement among populations takes into account the possibility that each ringed bird may arrive with its mate and family members, some of which may be unringed birds. This approach is possible because we routinely determine the size of social units during ring-reading observations. In this case, social status was recorded for 12 of the 15 immigrants (Table 13.1). Ten of these were recorded with a mate (four with ringed and six with unringed partners); one pair arrived with three goslings. Two others

Bird code	Original population	First seen on Solway	Duration of stay on Solway	Mate and family status on arrival
Yellow BHA	Greenland	Oct 1976	2 months[a]	With yellow BFU + 0 goslings
Yellow F Blue	Russia/Baltic	Winter 1986	5 years	With unringed partner + 0 goslings
Green A	Russia/Baltic	Oct 1988	-	With unringed partner + 3 goslings
White BVI	Greenland	Mar 1990	-[a]	Unknown
White AHA	Greenland	Nov 1991	-[b]	Arrived single
White BHT	Greenland	Dec 1991	3 months[a]	Arrived single; paired with unringed partner after arrival
White DTY	Greenland	Oct 1991	1 month[b]	With unringed partner + 0 goslings
White JST	Greenland	Feb 1996	3 months[a]	With JXX + 0 goslings
White FZC	Greenland	Oct 1994	8 months[a]	Arrived with unringed partner
White FUI	Greenland	Sept 1994	-	Unknown; dead in Svalbard
White LAZ	Greenland	Oct 1997	2 months[b]	With unringed partner + 0 goslings
White LTU	Greenland	Nov 1997	2 years 6 months	Arrived with unringed partner
White TZL	Greenland	Mar 1999	1 month[b]	Unknown

[a] Bird was resighted back in original range in a subsequent season.
[b] Bird was resighted back in original range that same season.

Table 13.1. *Status and fate of 15 individually marked geese that immigrated into the winter range of the Svalbard Barnacle Goose population. Most of these birds returned to their original population. Three of the 15 geese were observed in multiple years and were assumed to have summered in Svalbard.*

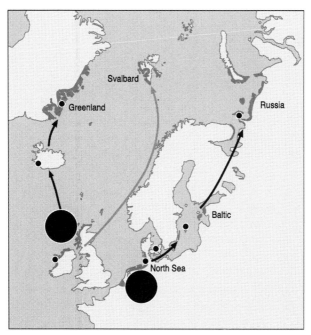

Figure 13.1. *Records of Barnacle Geese ringed in Svalbard and recovered in the range of other populations. Small circles represent 1–10 birds, large circles 28–34 birds. Sample size was 75 (68 birds with engraved plastic leg-rings and 7 birds with metal rings; Table 13.2).*

Destination	Females	Males	Total
Greenlandic winter range			
Islay, Scotland	16	16	34[a]
Ireland	3	3	6
Iceland	1	1	2
Russian/Baltic winter range			
Netherlands	14	14	28
Germany	1	1	2
Sweden	2	1	3
Bird was recorded in			
1 year and did not return	7	3	10
2+ years and did not return	5	10	16[a]
1 year and then returned	19	18	38[a]
2+ years and then returned	3	1	4
Returned in			
Same year	9	7	16
Next year	9	6	16[a]
Subsequent year	4	6	10
Age when moving (years)			
Range	0–15	0–14	0–15
Average	3.6	4.8	4.3
SD	3.1	3.4	3.4
Mate status at destination			
Alone/unpaired (# returned)	5	3	9[a]
With unringed mate	6	5	11
With ringed mate	5	5	10
With mate and goslings	1	1	3[a]
Unknown status	16	18	34

[a] Sex of one or two birds was not known.

Table 13.2. *Emigration statistics for marked birds from the Barnacle Goose population breeding in Svalbard detected in the wintering areas of one of the other populations. There were 68 fully documented cases and 7 records from additional sources (Ogilvie & Owen 1984 and WWT unpubl. data).*

apparently arrived alone. This amounted to nine additional immigrants that travelled with the 15 foreign-ringed birds during the study. While this method provides an estimation of the number of birds travelling with each marked bird, it does not estimate how many other unringed geese may have moved between populations.

A more elaborate method estimating rates of movement among populations is by multi-strata models, which account for population-specific survival and resighting probabilities. Analyses included the destination and origin of individuals, i.e. the population where the birds were marked. In this way, two different rates of movement between populations were calculated: one for birds moving to a *foreign* population, and the other for birds returning to their *original* population.[103] Numbers of emigrants and immigrants were extrapolated from the rates of movement by appropriate correction for the proportion of marked birds in the population.

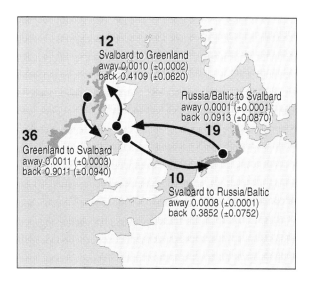

Figure 13.2. *Movements among three Barnacle Goose populations. For each population, emigration rate ('away') and return rate ('back') indicated as proportion of the birds (± SE), based on a Markovian transition model. Average number of birds moving annually indicated by bold figures. We have no evidence of movements between the Greenland and Russian/Baltic population.*

Figure 13.2 shows the results of these calculations. Movement rates were lowest for the Russian/Baltic birds (the probability of moving to the Svalbard population was 0.00014), and they moved five to eight times less than those travelling between the Svalbard and Greenland populations. Distance might play a role in these differences, although the rate of movement from the Svalbard population to the Russian/Baltic population was relatively high (0.0008). Similarly, the lack of records of birds moving between the Greenland and Russian/Baltic populations supports the idea that distance influenced movement rates.

Accounting for the average size of populations, 19 Russian/Baltic and 36 Greenland birds moved annually to the winter range of the Svalbard Barnacle Goose population, or a total of 55 immigrants. On the other hand, the Svalbard population provided 10 and 12 birds each year that emigrated to the Russian/Baltic and Greenland populations, respectively (Figure 13.2).

Most of the individuals that moved to another population stayed there for only a brief period before returning to the original range. Thirteen of the 15 foreign-marked immigrants were present in the Svalbard population during a single winter season and 10 of these were resighted back in the birds' original range (Table 13.1). In 42 of the 68 cases of emigration, the Svalbard-ringed geese returned in the same (38%), following (38%) or a subsequent year (24%; Table 13.2). It was particularly surprising to find that four birds that returned did so after an extended stay in another population (after 4, 5, 8 and 9 years). Of the 26 birds that did not return to the Svalbard population, 16 were seen in the foreign range in more than one year. Once emigrants spent a summer in the foreign population, they stayed for an average of 3.2 years (n = 13 for all populations combined).

Regarding the Svalbard population, for which the most extensive data are available, rates of movement to other populations are negligible in comparison to mortality rates. In fact, emigration to other populations is overcompensated for by immigration from other populations. Immigration exceeded emigration by an estimated 33 birds per year. We conclude that movements among populations have had little influence on the size of the Svalbard Barnacle Goose population during the course of this study.

Rates of movements among the Russian, Baltic and North Sea populations, which have a common wintering area, are probably higher than among populations with separate distributions. Van der Jeugd & Litvin (2006) documented the occurrence of 18 geese ringed on Gotland (two females, 16 males) that were resighted in another population during the summer breeding season. Ten birds were located in Arctic Russia, seven in other temperate breeding areas (one in Belgium, three in the Netherlands, three in Finland) and one in Svalbard. Their study also found that 'dispersers' were relatively large compared to all birds that were captured and measured (n = 3,693); seven out of 18 birds were among the largest 10% of their cohort.

By way of comparison, the rate of movement between Svalbard and Greenland populations of Pink-footed Geese has been estimated in a more robust mark–recapture analysis (Madsen *et al.* 2014). This study determined that the probability of exchange from Greenland to Svalbard was 0.00071 and from Svalbard to Greenland 0.00076. Like the Barnacle Geese, the stray Pink-footed Geese returned to their original range; for example, 21 of the 32 birds ringed in the Svalbard range that moved to the Greenland range were resighted in their original range in the next season; the fates of the other 11 were unknown.

Immigration into the Svalbard population

Even though minor exchange of individuals among populations may not greatly affect the demographic structure or growth of the populations, it may have genetic consequences. For example, even moderate exchange rates may preclude genetic differentiation among populations and speciation. To elucidate this question we analysed movements among Barnacle Goose populations in terms of exchange of genetic material; or in other words, the rate of gene flow (Hewitt & Butlin 1997). Below we estimate gene flow from calculated immigration rates into the Svalbard population and compare this value with published values at which flow speciation might occur.

We estimated that on average 55 foreign birds immigrated annually into the Svalbard population (see above), and only 10 of these took up residence in the population.[104] In other words, the majority of foreign birds were present for only a short time during the non-breeding season and the opportunity for gene flow would seem minimal. By assuming that the long-staying immigrants reproduce at rates similar to others we can estimate the number of foreign genomes that may have been added to the Svalbard study population. Barnacle Geese are capable of breeding in their second year (van der Jeugd & Larsson 1998), and on average 17% of breeding-age females succeed in returning to the wintering area with an average of two surviving juveniles (Chapter 9). Annual survival rate for the youngest age classes is approximately 0.90 (Chapter 9). Given that immigrants had three breeding seasons in the foreign population (see above), the 10 immigrants would produce eight goslings that would reach two years of age, the age of first potential breeding.[105] This amounts to eight additional sets of foreign genomes via reproduction that would be incorporated into the Svalbard population each year.

The influx of foreign genes at this rate would, no doubt, make an impact over long periods of time. It has been estimated that as few as 5–16 immigrants per generation inhibit the formation of a new species (Slatkin 1987, Porter & Johnson 2002), although speciation also depends on factors other than gene flow (e.g. MacArthur 1972). The estimate for the Svalbard Barnacle Goose population of about 17 'effective' immigrants per generation[106] may be why no apparent subspeciation has occurred in the Barnacle Goose.

Genetic structure of populations

Jonker *et al.* (2013) used molecular genetic techniques and analysed DNA from blood and tissue samples from all five Barnacle Goose populations. They showed that the populations were genetically differentiated and that the genetic distance between the Svalbard population and the Russian, Swedish (Baltic) and Dutch (North Sea) populations was larger than among the last three populations, with the Russian population being the closest of these three to the Svalbard population. These results suggest that the temperate Baltic and North Sea populations originated from the Arctic Russian population. Moreover, the result that the genetic distance between the Russian and Baltic populations was the smallest of all pairwise comparisons indicates that most exchange of individuals has occurred between these populations. Jonker *et al.* (2013) also argued that the presence of 'linkage disequilibrium' in the analysed genetic material implied that individuals have moved from one population to another over the past few generations at a rate generally higher than what can be detected from ring recoveries.

From observations of movements of ringed birds as well as from the molecular genetic analyses it can be concluded that the Barnacle Goose populations are genetically differentiated to a varying extent but also that gene flow has occurred between the populations. We anticipate that mixing among populations will become more frequent in future because all five Barnacle Goose populations are increasing in number and expanding their ranges. If breeding distributions continue to meld together, the present genetic structure may disappear quite quickly or reappear in another way.

Pioneering tendencies or accidents during migration?

Do individuals change populations because they are different from others and for some reason have a higher propensity to move, or are they victims of 'bad luck' and lose their way during migration? A pioneering tendency is commonly attributed to young animals that have not yet become established at particular sites (Swingland & Greenwood 1984). We have argued that younger birds tend to be the pioneers in exploring alternative sites – at least that was the case for within-population movements (Chapter 12). Indeed, 48% of the 68 birds described in this chapter that were resighted in another population were less than three years of age. However, the other 52% of confirmed emigrants were between three and 15 years of age. The average age at which birds were detected in the foreign location was 4.2 years, which is beyond the usual age for exploratory behaviour of younger animals. Although we cannot exclude the possibility of an age-related tendency to move, the age comparison and the observation that most birds returned to their 'home' population may indicate that many of these birds lost their way, for example during severe weather. Cold weather movements by waterfowl from mainland Europe to Britain has been well documented (Ridgill & Fox 1990, Madsen *et al.* 2014). Further work is required to find out whether younger birds have a greater tendency to shift to and then remain in a different population (*sensu* van der Jeugd 2013).

Summary and conclusions

In this chapter we argue that immigration and emigration have had a negligible impact on the demographic structure and growth of the Svalbard population of Barnacle Geese. It seems that most movements into foreign goose populations were by singles, pairs and small groups that had lost their way. Based on transition probabilities we estimated that each year 55 foreign birds immigrated into the Svalbard population and 22 Svalbard birds emigrated to another population. However, evidence from resighting ringed birds suggests that the stay in the foreign population was brief; only 18.8% of the immigrants and 29.4% of emigrants remained in the foreign population for more than one year. Most birds were observed back in their original population in the same or next year. The foreign birds that remained in the Svalbard study population produced eight 'foreign-derived' offspring each year. Molecular genetic analysis showed that the Barnacle Goose populations were genetically differentiated. We anticipate that mixing among populations will become more frequent in future because all five Barnacle Goose populations are increasing in number and expanding their winter and breeding ranges.

Statistical analyses

[103] The multi-strata model included the following parameters: survival and resighting probabilities (assumed constant for each population), and movement rates among populations (the origin of birds was treated as an additional group variable). To reduce the number of parameters, movements between the Russian/Baltic and Greenland population were not differentiated. Encounter histories for the Greenland and Russian/Baltic population were simulated on the basis of estimates of survival and resighting probabilities, and on the number of ringed birds throughout the study period (S. Percival pers. comm., Percival 1991, Ebbinge *et al.* 1991, van der Jeugd & Larsson 1998). An estimate of the number of Gotland-ringed birds was taken from maximum values in Appendix 1 in Larsson *et al.* (1998). We used the formula of Larsson & van der Jeugd (1998) to calculate the Baltic population size based on the annual number of breeding pairs (Larsson & Forslund 1994). The Greenland population size was assumed to be midway between most recent census values reported in Ogilvie *et al.* (1999). An estimate of the number of ringed birds was calculated by assuming an average annual survival of 0.85 between 1984 and 1999 (S. Percival pers. comm.). Russian population size estimates were based on midwinter surveys (B. Spaans pers. comm.).

[104] Assuming 55 immigrants per year, as calculated from the multi-strata model, of which 18.8% stayed in the Svalbard population, means that each year $55 \times 0.188 = 10$ birds would have taken up residence in the population.

[105] Calculation of annual production of offspring by immigrants in the Svalbard population: 10 (immigrants arriving per year) × 3 (number of years present) × 0.17 (proportion of potential breeders that are successful) × 2.0 (average brood size in winter) × 0.90 (first year survival) × 0.90 (second year survival) = 8.3 offspring per year.

[106] Calculation of immigrants per generation: 10 immigrants per year × 10 years (generation time) = 100 immigrants per generation. On average, only 17% of all Barnacle Geese reproduce successfully per year, which means that the number of effective immigrants is reduced to 17.

CHAPTER 14

Population dynamics

Population dynamics is the study of population behaviour, asking how and why numbers change over time and over the landscape. We define wildlife populations based on the biology of the animal together with geographical and sometimes political jurisdictions (e.g. Svalbard or Baltic Barnacle Goose populations). Often populations are composed of multiple subpopulations inhabiting discrete habitat patches at, for example, the breeding grounds (colonies) or other sites visited during the annual cycle (*sensu* Levins 1969). Growth rates of the population as a whole will be affected by the periodic establishment and extinction of subpopulations that behave in response to possibly unique conditions within the geographic range. To interpret or predict population behaviour it is necessary to consider processes (e.g. competition) which may give rise to density-dependent effects and regulate population size (*sensu* Lack 1954). Density-dependence occurs when resources like food or nest sites are limited and competition leads to unequal allocation among individuals, resulting in reduced population growth. This population regulation becomes stronger when numbers increase toward 'carrying capacity' of habitats, and the effects weaken when numbers decline (Figure 14.1). Thus, there is a negative feedback between population density and population growth. Because Barnacle Goose populations have increased to levels that are unprecedented in historic times, we expected this feedback to influence the rate of population growth. Other important factors act independently of population size or density; for example, severe storms during migration or other extreme weather conditions may influence the survival of juveniles and adults. Predators may also greatly affect numbers at a local scale by taking eggs, goslings or full-grown geese, and perhaps even more so by reducing the space where geese can safely roam. Activities by predators may be linked to density but often without regulating overall population size.

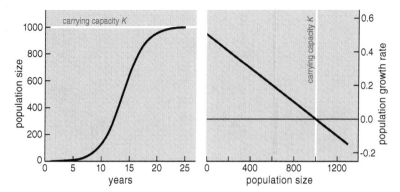

Figure 14.1. *(A) Theoretical population that follows a logistic growth pattern. Maximal numbers are set by the environment (carrying capacity, K). (B) Population growth rate declines with increasing numbers. The maximal growth rate is at the intercept with the y-axis (0.5 in this case). Numbers above carrying capacity would cause a negative growth rate, or numbers decline.*

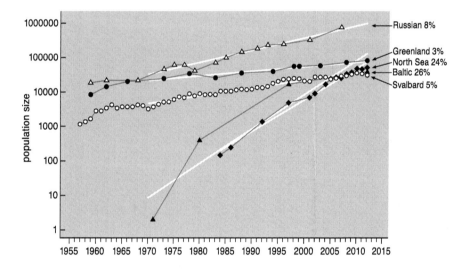

Figure 14.2. *Pattern of population growth of the five European Barnacle Goose populations. Sizes are plotted on a log-scale, with slopes indicating growth rates since 1975 (North Sea population from 1984). Note the larger Russian and Greenland populations in comparison to the smaller Svalbard, Baltic and North Sea populations. Derived from Ganter et al. (1999), Larsson & van der Jeugd (1998), Ogilvie et al. (1999), Owen & Black (1999), Ouweneel (2001), Meininger (2002), WWT (2013), www.sovon.nl, and H. van der Jeugd, K. Koffijberg, E. C. Rees and P. Cranswick, pers. comm.*

Figure 14.2 depicts the trends of numbers for Barnacle Goose populations, all of which exhibited rapid growth over time. Average growth rates of the populations, indicated in the figure, were derived from the slope of the numbers against time (see below). The Greenland population increased at the slowest rate (3% annually from the late 1950s to 2012), which is considerably lower than the performance of the Svalbard and Russian populations (5 and

8%, respectively). The recently developed Baltic and North Sea populations increased at extremely high rates (26 and 24%, respectively), and both populations have overtaken the Svalbard population in size.

In this chapter we describe how the combined effect of the density-dependence and other processes has shaped our two study populations. We begin by describing the methods we used to track changes in numbers over time, focusing on the Svalbard population.

The Svalbard population

The clamour of large flocks circling overhead marked the return of the geese to the Solway Firth on the border of Scotland and England. Each year refuge managers, bird watchers and farmers were keen to find out how many more there would be, how well they had bred and which pastures and crops they would find. In addition to flock counts, two additional parameters were recorded annually, namely average brood size and the percentage of juveniles in the winter flocks. These observations were made when the majority of the population was present in front of the hides and observation towers at Eastpark Farm (WWT Caerlaverock) at the beginning of the wintering period. Using this basic data a range of population parameters were calculated, including age composition, measures of productivity, survival and rates of population growth (Box 14.1). For example, Figure 14.3 shows that the number of juveniles surpassed the number of deaths in 25 of 40 years. This helps to explain that the population growth rate was the result of the net effect of births and deaths rather than, for example, immigration of foreign birds into the population. In Chapter 13 we established that immigration and emigration have had a negligible effect on the size and demographic structure of the Svalbard Barnacle Goose population.

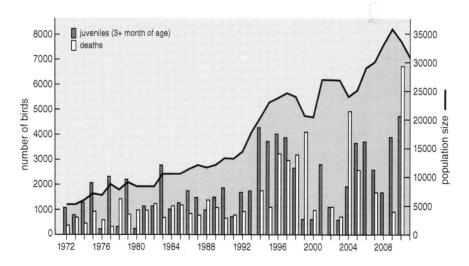

Figure 14.3. *Number of juveniles in the winter flocks compared to the number of deaths since the previous winter in the Svalbard Barnacle Goose population (1973–2012). In most years the number of juveniles exceeded deaths, which resulted in an increase in population size (indicated by dark line). Updated from Owen (1984), Pettifor et al. (1998).*

BOX 14.1 POPULATION PARAMETERS

Population parameters derived from winter assessments of total population size N_t, proportion of juveniles PJ_t, and brood size BS_t, in which t stands for time.

(i) Number of juveniles J_t in winter t is calculated as

$$J_t = PJ_t \times N_t$$

(ii) Number of one-year-old birds, yearlings Y_t, can be calculated as

$$Y_t = J_{t-1} \times GYS$$

where GYS is the survival of goslings from their first to second winter when they are yearlings. For GYS we used the average value of 0.895 (SE 0.0065) from six cohorts of individually marked yearlings (Chapter 9).

(iii) Number of potential breeders A_t is equivalent to the number of adults in the population, defined here as the number of birds that are at least two years old. The number of adults is calculated as

$$A_t = N_t - (J_t + Y_t)$$

(iv) Number of successful breeders SB_t (having at least one juvenile when returning to the wintering grounds) can be calculated as

$$SB_t = 2 \times (J_t / BS_t)$$

(v) Proportion breeding PB_t is the proportion of potential breeders that successfully bred. This is calculated as

$$PB_t = SB_t / A_t$$

(vi) Annual survival rate S_t can be estimated as

$$S_t = (N_t - J_t) / N_{t-1}$$

Survival rates estimated in this manner will be underestimated in some years and overestimated in others, but it is assumed that these errors are self-compensatory over time (e.g. Owen 1982, 1984). When available, capture–mark–recapture estimates of survival are preferable (Chapters 6, 9, 10).

(vii) Number of deaths (M_t mortality) between the previous and current year can be estimated as

$$M_t = N_{t-1} - (N_t - J_t)$$

(viii) The finite rate of increase λ is the factor by which population size changes, and can be calculated as

$$\lambda = N_t / N_{t-1}$$

(ix) It is often more convenient to use the natural log of λ, by convention called the population growth rate r:

$$r = \ln (\lambda)$$

Another way to estimate the long-term growth rate r is from the slope of the regression of population size (on a log-scale) against year (Sibly & Hone 2003). Population growth rates, λ and r, are often estimated from age-structured population models, which describe numbers, survival and reproduction of organisms

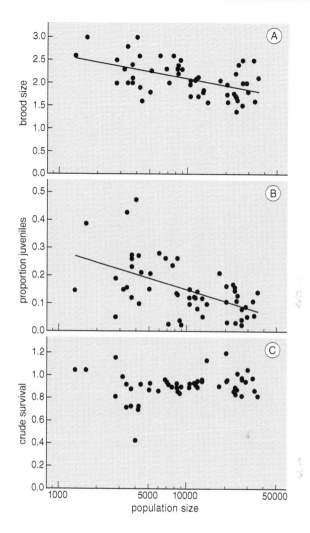

Figure 14.4. *Population parameters from 1959 to 2012 in relation to the size of the Svalbard Barnacle Goose population in the previous year. Brood size (A) and proportion of juveniles in the winter flocks (B) were negatively related to population size. Crude annual survival rate (C) did not show an association with population size. Updated from Owen (1984), Pettifor et al. (1998).*

Density-dependence

There have been profound changes in the Svalbard Barnacle Goose population structure over the course of the study. For example, the proportion of juveniles in the winter flocks declined by more than 50%, which resulted from the steady decrease in both the proportion of successful pairs and average brood size as the population increased in size (Figure 14.4).[107] These negative relationships are clear signs of density-dependent effects operating in the population. It appears that growing numbers have a strong negative effect on reproductive success, but this may be partly compensated for by steady or slightly increasing adult survival rates (Figure 14.4C).

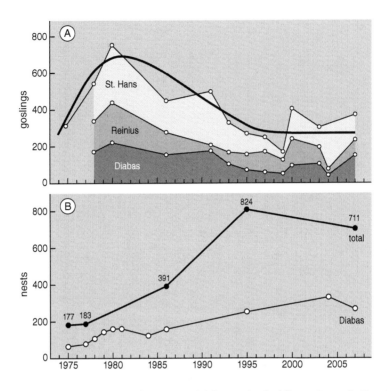

Figure 14.5. *Gosling production (A) and nest counts (B) for Nordenskiöldkysten Barnacle Geese from 1975 to 1997. After an initial increase, gosling numbers declined (panel A) in spite of a continued increase in breeding pairs (nests, panel B). Nest counts were made post-hatch and ground counts of brood-rearing areas were made while hiking the coastline (see Figure 2.4, 2.5). The most intensively studied colony, Diabasøya, was visited more often than the other colonies in the area (lower plot on panel B). In 1980, 75% of paired adults were associated with young during the brood-rearing period, but by 1995 this value declined to only 16%. Derived from Drent et al. (1998).*

Density-dependence was noticeable at the colony level. Colonies on offshore islands along Nordenskiöldkysten produced maximal numbers of goslings only during a brief period (late 1970s–early 1980s). This was in spite of a continuous increase in the local population from 1,200 to 4,600 geese that competed for territories in local colonies between 1976 and 2003, respectively. For example, in 1980 about 775 goslings were produced from 200 nests and 15 years later only 250 goslings survived from 824 nests (Figure 14.5). The drop in reproductive success may have been due to competition on the brood-rearing areas (Chapter 8), perhaps exacerbated by events during the pre-incubation or incubation periods.

Weather during summer

Goose biologists agree that summer weather conditions are a primary determinant of the growth rate of Arctic goose populations (Barry 1962, Newton 1977, Owen 1980a, Prop & de Vries 1993, Boyd & Madsen 1997). If cold weather persists into the breeding season, the Arctic summer is delayed and opportunity for successful breeding is reduced. Warmer weather in the Arctic initiates primary production in the tundra habitat. The effect of

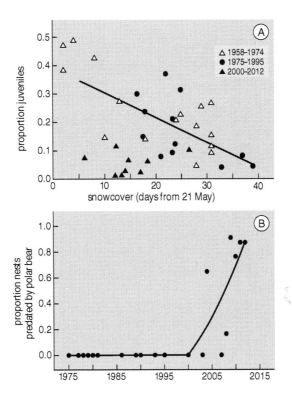

Figure 14.6. *(A) Relationship between the percentage of goslings in moulting flocks, Nordenskiöldkysten, and the duration of snow cover (number of days from 21 May when at least 75% of the tundra was covered by snow). Observations for 1975–1995 and 2000–2012 are given separately. For comparison, earlier data collected in the wintering area in 1958–1974 are shown (Owen & Norderhaug 1977). The regression line is for the 1975–1995 period. The winter assessments (1958–1974) do not differ from this regression.[110] After 2000 the significant relationship between juvenile proportions and snow cover disappeared.[110] Updated from Drent & Prop (2008). (B) Proportion of nests predated by Polar Bears in the main study colony in Svalbard. Polar Bears visited this colony for the first time in 2004. The number of nests during the years of Polar Bear raids ranged from 337 to 522.*

favourable weather conditions becomes obvious when comparing the annual proportions of juveniles in the brood-rearing flocks in Svalbard and the extent of snow cover in the early breeding period (Figure 14.6A). For the years 1958–1995, length of snow cover on the Svalbard tundra was an extremely important determinant of gosling production.[108] As a consequence, annual population growth rate was closely linked to snow conditions on the breeding grounds.[109] However, in recent years as Arctic temperatures have increased this pattern has changed. When considering Figure 14.6A carefully it is striking that the initial close relationship between the proportion of juveniles and snow cover was lost from the onset of the 21st century (the most recent years are depicted by filled triangles in Figure 14.6A).[110] The explanation for the change in this relationship will be discussed in the next section under a different theme because the warming climate has added an additional predator to the landscape where the geese breed. With warming temperatures significant parts of the ocean around the Svalbard archipelago no longer remain frozen in summer.

Predation pressure

The cause of the sudden fall in goose productivity is all too clear and can be attributed to the appearance of Polar Bears in the region. Rather than moving in summer to the ice-covered hunting grounds east and north of Svalbard, Polar Bears tended to stay on land in increasing numbers (Drent & Prop 2008). This was a dramatic change for the Barnacle Goose (and several other bird species), and it followed on from a rapid decline in the extent of summer sea ice and a recovery of the bear population after it gained protection in 1973. This shift in Polar Bear summer distribution coincides with developments observed in other regions around the North Pole (Rockwell & Gormezano 2009, Stirling 2011, Iverson *et al.* 2014). Polar Bears strolling along the west coast of Svalbard have no difficulties in swimming to goose colonies on offshore islands. While moving across the islands at their slow pace, the bears find one goose nest after the other and can eat the contents within 1.5 minutes. When bears arrive in time, i.e. before the eggs have hatched, the predation rate can be high, amounting to over 90% of the nests (Figure 14.5B). The bears on Nordenskiöldkysten have taken about 200 nests per visit in recent years, which is not many in terms of the entire breeding population. However, it is the recurring presence of Polar Bears in the breeding colony that causes trouble for the geese. In 2012, for example, the first bear appeared on 9 June, which was the date that most of the geese had settled and started incubation. During two visits on two consecutive days it emptied 265 nests (including those of some Eiders and Glaucous Gulls). A week later three bears in succession visited the colony, together predating 261 nests. By the end of June two more polar bears came to collect the eggs from 85 nests, and in July two Polar Bears took all the remaining eggs.

Barnacle Geese seem very capable of coping with the constant threat of their other main predator, the Arctic Fox, by choosing nests sites that are inaccessible (islands, rock stacks, cliffs) and by feeding in close proximity to lakes where they can retreat when a fox approaches during the moulting and brood-rearing periods. However, the recent appearance of Polar Bears in the world of the Barnacle Geese shows that they are vulnerable to new predators. Production of goslings on Nordenskiöldkysten has dropped severely and the geese will have to find alternative places for nesting that are safe from Polar Bears and foxes. Additional sites with steep cliffs, their original breeding habitat (Chapter 2), can be found in the interior of Svalbard and may be a suitable alternative.

The Baltic population

Density-dependence

Although emigration and immigration between the temperate Baltic population and the Arctic populations were detected during the study the rates were so low that the overall net effect on the growth of the Baltic population was negligible (Chapter 13). The increase in numbers and the breeding range expansion of the Baltic Barnacle Goose population were instead best explained by the net effect of births and deaths as well as by dispersal to and colonisation of new areas within the Baltic region (Figure 2.1, 14.7).

The pattern of very rapid initial growth, subsequent stabilising and in some cases even decline in numbers was repeatedly observed both in colonies and on regional levels within

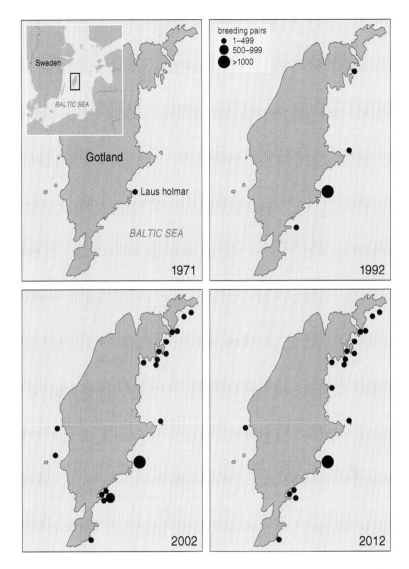

Figure 14.7. *Distribution of Barnacle Geese colonies on Gotland in 1971, 1992, 2002 and 2012.*

the Baltic (Figure 14.8). In the main study colony, the Laus holmar colony, the number of breeding pairs increased at a phenomenal rate from just a single breeding pair in 1971 to 630 pairs 15 years later, and to 2,450 pairs in 2001. The average annual growth rate during the first 15 years was 64% whereas during the second 15-year period the average annual growth rate was 8%. The south-eastern coast of Gotland was first colonised in 1987 and the average annual growth rate there during the following 15 years was again very high, at 59%. The same pattern was observed along other coastal stretches of Gotland and Öland (Sweden), and in west Estonia, Helsinki area (Finland) and on Saltholm (Denmark) (Leito 2011, Mortensen 2011, Väänänen *et al.* 2011; Figure 14.8).

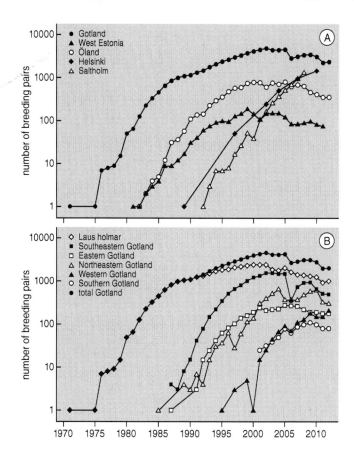

Figure 14.8. *(A) Growth of Barnacle Goose colonies within five surveyed areas in the Baltic region. Numbers of breeding pairs are plotted on a log-scale. Data from West Estonia, Helsinki area (Finland) and Saltholm (Denmark) were obtained from Leito (2011), Väänänen et al. (2011) and Mortensen (2011), respectively. In 2012 we estimated that a few thousand pairs also bred outside the five surveyed regions in Sweden, Finland and Denmark in coastal and urban areas. (B) Growth of Barnacle Goose colonies in six coastal areas on Gotland. The patterns of initial very rapid growth and subsequent reductions in growth rates, or even declines in numbers, were repeatedly observed on different geographical scales and indicate the presence of density-dependent population regulation.*

In the Laus holmar colony, the production of fledged young per breeding pair declined in a density-dependent manner as the number of breeding pairs increased (Figure 14.9). Average clutch size varied among years but did not decline with increasing number of pairs. This pattern was also observed in other colonies on Gotland. Thus, the drop in gosling production could not be explained by, for example, increased competition for food during the pre-laying period. Instead, the decline was closely linked to changes in gosling survival between hatching and fledging. This finding is in agreement with the hypothesis that the availability of high-quality brood-rearing sites in the vicinity of the nesting inlands strongly affects gosling growth and survival. Increased crowding on brood-rearing sites will make goslings more vulnerable to starvation and predation.

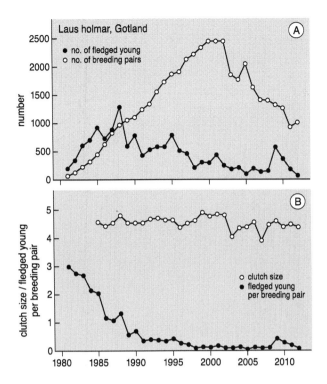

Figure 14.9. *(A) Number of breeding pairs and number of fledged young in the main Baltic study colony on the island group Laus holmar off the east coast of Gotland. The number of breeding pairs increased up to 2002. Predation on adults and nests by Red Fox was detected in 2003 and 2004 on the largest island and in 2011 on the two smaller islands. No fox predation was detected in any other year. From the beginning of the 2000s predation by White-tailed Eagles on nesting adult Barnacle Geese increased considerably. (B) The number of fledged young per breeding pair declined in a density-dependent manner as number of breeding pairs increased up to 2002. However, the number of fledged young per breeding pair did not increase after 2002 when the number of breeding pairs started to decrease. The lack of upward response is most likely due to increased predation and disturbance by White-tailed Eagles. There was no relationship between average clutch size and population density.*

The strong density-dependent effects on gosling survival led to situations where small newly established colonies actually produced more fledged goslings than large well-established colonies. For example, in 2001, 30 years after colonisation, the Laus holmar colony produced only 440 fledged young from 2,450 breeding pairs compared to a production of 505 fledged young from 305 breeding pairs at younger colonies on north-eastern Gotland. When considering all colonies on Gotland in 2001, the oldest colony hosted 56% of the total number of breeding pairs but produced only 25% of the fledged young.

Predation pressure

Although density-dependent effects on gosling production were a major cause of the reduction in colony growth rates as number of breeding pairs increased, other effects were also observed. Since the middle of the 1990s the White-tailed Eagle has increased in numbers in the Baltic region from being almost extinct there during the 1970s. The successful recovery of breeding White-tailed Eagles on Gotland and in the central Baltic region has led to increased predation on nesting adults and large juveniles in some of the largest Barnacle Goose colonies. Since about 2005 we estimated that one or several White-tailed Eagles hunted in the Laus holmar colony almost daily during the incubation and brood-rearing period of the geese. On some occasions up to eight different eagles simultaneously hunted geese and other waterfowl within the large 0.6km² colony. During such chaotic events Herring Gulls and Great Black-backed Gulls often took advantage of the situation and searched for eggs in unattended goose nests. The decline of the number of breeding pairs in the Laus holmar colony and in some other colonies on Gotland and Öland since this time can therefore partly be ascribed to increased predation and disturbance by White-tailed Eagles (Figure 14.9). The return of this avian predator may also have encouraged Barnacle Geese in the Baltic region to utilise new nesting habitats, such as forested islands and urban habitats (Väänänen *et al.* 2011). Similar trends of breeding in less open habitats have also been observed in Baltic breeding Eiders, a species that often breeds intermingled with Barnacle Geese.

Population range expansions

In Chapter 12 we showed that as density-dependent effects were realised at well-established colonies, natal dispersal rates increased and gave rise to numerous new colonies. In both the Baltic region and in the Svalbard archipelago, multiple colonisation events have taken place and there are now a large number of colonies spread along the coasts (Figures 1.7, 14.7).

Figure 14.10 depicts our working hypothesis for the continued increase of Barnacle Goose populations. This includes the prospecting of new areas and establishment of additional colonies that are initially free from density-related processes. After an initial period of learning how best to exploit local resources while avoiding local predation pressure (Black 1998a, Loonen *et al.* 1998, Tombre *et al.* 1998) the overall population is fuelled by the contribution of juveniles from each new colony, whereas periods of stability are due to the limitations brought on by harsh weather and by competition for food on the breeding grounds affecting survival and reproductive success. Based on this scenario, therefore, the process of dispersal and successful colonisation of new habitats is the main mechanism behind the increasing Barnacle Goose populations.

These ideas are based on the assumption that colonies behave very differently, but that many of them will eventually experience density-dependent limitations on reproduction. We observed that colonies, after being established, developed through the following stages: (i) an initial lag between colony establishment and the first successful breeding, (ii) a rapid growth in numbers and high reproductive success, (iii) stabilising of reproductive success, while numbers are still increasing, and (iv) a drop in reproductive success, and a local population that stabilises or decreases in numbers. The most productive period of a colony seemed to last for only 15–20 years, after which reproductive rates began to decline.

It follows that individuals from different colonies are contributing recruits to the larger population at different rates, findings that were obtained in a parallel study of a Black Brant 'metapopulation' in Alaska (Sedinger *et al.* 2002). Therefore, considering information from

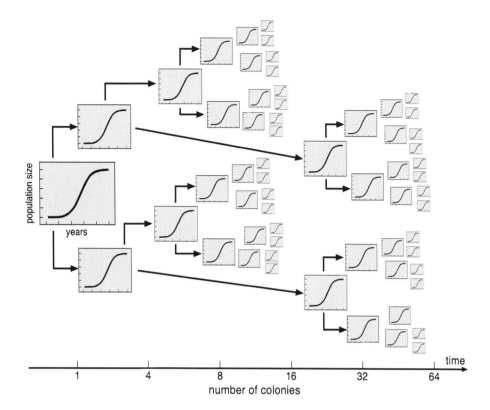

Figure 14.10. *Schematic representation of exponential growth in a goose population. Several hypothetical colonies are established at different times. Each subpopulation eventually suffers a density-dependent decrease in growth rate, leading to an increasing rate of dispersal and colonisation of new areas. Each new colony gives rise to additional subpopulations. An increasing number of subpopulations maintain an exponential growth, although older subpopulations show no growth.*

only a few subpopulations may lead to the description of spurious population responses (Ranta *et al.* 1995, Bonsall *et al.* 2002). This is especially relevant for migratory animals that range across political boundaries where priorities for land use and habitat management vary and different hunting regulations occur.

Summary and conclusions

The chapter began with a review of the population parameters that can be generated with basic field assessments of the proportion of juveniles and brood sizes in wintering flocks. Winter counts indicate that all five Barnacle Goose populations increased annually by 3–22%. For example, the Svalbard population increased in size in 25 of 40 years, when juvenile numbers surpassed adult deaths. Both study populations were affected by strong density-dependent effects, climate and predation pressure. Successful reproduction and growth rates of the Svalbard population were strongly linked to the length of snow cover on the breeding grounds – late snowmelt led to low breeding success. Once newly formed colonies become established, geese may need time to learn about the intricacies of new habitats and predation risks. Gosling production was high during the first few decades after colony establishment, after which production of young declined. Therefore, as time went on most goslings were produced in newer colonies that were not yet hindered by competition for limited resources. We suggest that the rate of population increase is also affected by the probability of dispersal, the availability of new breeding sites, and the period of learning the intricacies of new habitats. The duration of phases of population increases and the subsequent periods of stability will be determined by density-dependent and other factors operating at the level of subpopulations.

Remarkably, performance of both study populations was recently affected by a steep increase in predation rates. In Svalbard this was because an increasing number of Polar Bears are searching for food on land (due to changes in sea ice conditions), and in the Baltic because of the recovery of the White-tailed Eagle breeding population. Predation effects were particularly strong on a local scale, and future observations will tell to what extent the total populations will be affected.

Statistical analyses

[107] Regression statistics of brood size and proportion of juveniles in the winter flocks, and crude annual survival in relation to population size: $t = 6.53$, $P < 0.0001$; $t = 1.99$, $P = 0.05$; $t = -0.04$, NS; $n = 54$. See Figure 14.4.

[108] Regression of proportion of juveniles in moulting flocks related to snow cover in Svalbard (number of days with snow cover above 75%) for 1975–1995: $y = 0.390 - 0.0087x$, $t = -2.25$, $P < 0.05$ (Figure 14.6A). There was no difference in the relationship between the proportion of juveniles (determined in wintering flocks) and snow cover in Svalbard in 1975–1995 (based on summer data) and in 1958–1974 (Owen & Norderhaug 1977): $F_{1,26} = 0.247$, NS. During the years 2000–2012 the relationship between juvenile proportions in moulting flocks and snow cover was not significant: $F_{1,7} = 0.445$, NS.

[109] Multiple regression of population growth rate ($n = 42$) in relation to (i) snow cover in Svalbard ($F_{1,39} = 25.2$, $P < 0.0005$, partial $r2 = 0.392$), and (ii) population size in the previous year ($F_{1,39} = 6.3$, $P < 0.025$, partial $r2 = 0.139$). Population growth rate was unrelated to temperature during spring staging ($F_{2,37} = 0.9$, NS).

[110] First, the proportion of juveniles was regressed against snow cover and local population size (based on data for 1975–1995). This model was used to calculate residuals of the regression for the years including most recent years (1975–2012). The residuals showed a steep decline during the second part of the study period ($F_{2,17} = 3.61$, $P < 0.05$; $Beta_{year} = 6.82$, $Beta_{year-squared} = -7.26$). See Figure 14.6A.

CHAPTER 15

Conservation and agriculture

Fifty years ago, wintering geese were predominantly found in wetland marshes and coastal habitats. Due to the abundance of agricultural crops and refuges today, goose distribution is now more widespread, bringing them into closer contact with people. While this proximity is celebrated by some, it is a problem for those trying to make a living from growing the crops. In her book *Wildfowl and Man*, Janet Kear (1990) suggested that conflict between geese and farmers is as old as agriculture itself, and referred to documentation of geese exploiting crops over 3,000 years ago in northern Africa. She also charted the phenomenon of potato and turnip eating by British waterfowl beginning in the 1800s. The discovery of this food could be traced to years of particularly harsh weather when traditional food sources were covered with snow. The use of the new food caught on in subsequent years, and today many geese still regularly make long inland flights from their coastal roosts in search of these nutritious crops.

For the entire year the Baltic and North Sea Barnacle Goose populations use agricultural fields and saltmarshes grazed by cattle and sheep (van der Jeugd *et al.* 2001, 2009). In winter, the Svalbard population also makes use of well-managed agricultural habitats in Britain, but in spring, in Norway, the shift to agriculture was gradual. Prior to about 1995 most of the Svalbard population still made use of salt-tolerant vegetation on the remote, sea-swept islands of Helgeland (Chapter 2). In the first sections of this chapter we describe the shift from these more natural habitats to agricultural landscapes and the consequence this has had on the birds' foraging performance and reproduction. We then discuss the notable conservation efforts by private conservation organisations and government agencies that attempt to reduce losses to individual farmers.

The shift to agricultural habitats

Over the course of our study in Helgeland, we observed a sizeable shift away from the traditional islands to agricultural habitats located on larger islands closer to the mainland (Figure 15.1). During the 1980s there was a big change in the composition of vegetation communities on the traditional outer island habitats (Gullestad *et al.* 1984) resulting in a reduced carrying capacity of the area (Prop *et al.* 1998). The vegetation changed because the traditional management of hay making and grazing by sheep and cattle was given up, and

Figure 15.1. *Spring staging area in Helgeland, Norway (65°N). Geese acquire different amounts of fat and protein stores from the used habitats. (A, B) Barnacle Geese foraging on agricultural fields at Tenna/ Herøy, large islands close to the mainland (both photos by Paul Shimmings). (C) The outer islands of Lånan located 30km from the mainland. Overgrown hay meadows are amidst barns and Eider houses that were formerly used by island crofters (Jouke Prop).*

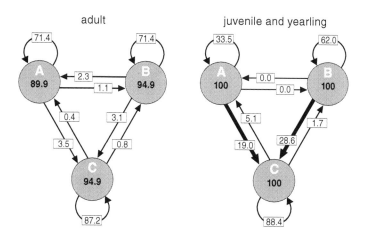

Figure 15.2. *Annual movement (arrows) and survival (values inside circles) probabilities of 2,212 Barnacle Geese travelling between island types (straight arrows) and choosing site fidelity (curved arrows) in the spring staging area, Helgeland, Norway (1987–1993). There was a net movement from traditional islands (A and B) to newly created agricultural islands (C) for both adults and subadults. Site fidelity was particularly low for subadults on traditional outer islands that had been abandoned by crofters (A). Survival, measured from one spring to the next, was not significantly different among staging islands. Values do not add up to 100% because a proportion of the birds moved to 'unknown locations'. Model results are given in Appendix 1 (Table 8).*

the higher parts of the islands were overgrown by tall vegetation (Chapter 2). The change in vegetation on the traditional islands had a dramatic effect on the birds' distribution as they became solely dependent on low-elevation saltmarshes. As the goose population continued to grow and maximal numbers on saltmarshes had been reached, an increasing number of geese moved closer to the mainland where agricultural fields were intensively managed (Black *et al.* 1991, Prop *et al.* 1998). The younger cohorts in the population, which had not established successful foraging routines at the original sites (Chapter 8), initiated the shift to the agricultural areas (Figure 15.2). Many older birds, on the other hand, remained faithful to their traditional outer island habitats; whereas the probability of return to the same island in the next year was 71.4% for adults, it was only 33.5% for young birds. This finding parallels the discovery of new breeding sites by younger birds described in Chapter 12. Farmers at the newly discovered areas did not initially tolerate the geese and intensive scaring (hazing) campaigns were initiated, which encouraged a further shift in bird distribution.

Opportunities to find suitable foraging areas for the rapidly growing numbers of geese during the last decades of the 20th century were limited within the boundaries of Helgeland. When numbers passed 20,000, excess individuals were accommodated in the rural area of Vesterålen, 300km northwards (Shimmings & Isaksen 2003, 2013, Tombre *et al.* 2013a). Figure 15.3 depicts this northern shift in distribution from the original staging range in Helgeland. In addition to shifts in habitat structure and carrying capacity problems within the traditional range, climate change seemed to have played a crucial role in the distribution of Barnacle Geese. From 1975 onwards, the average temperature during the goose staging period (21 April – 15 May) increased by 2°C. Within the traditional range, the number of days warm enough for grass growth more than doubled during this period, and in the

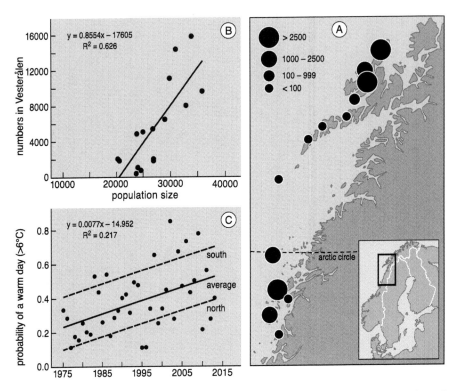

Figure 15.3. *Distribution of Barnacle Geese in spring staging areas on the Norwegian coast. Indicated is the original range of the birds in the 1970s, the extension in the 1980s, and the subsequent further colonisation to the north (A). (B) Increasing numbers reaching the northernmost Vesterålen region (Shimmings & Isaksen 2013, Tombre et al. 2013a) in relation to the size of the flyway population.[111] (C) The trend in increasing temperatures during the goose staging period from 1975 onwards, portrayed here as the probability of a day warmer than the lower threshold of plant growth (6°C).[112] Dashed lines indicate trends for a northern (69°N) station in Vesterålen and a southern (65°N) station close to Helgeland. Numbers reaching carrying capacity in one area and a warming climate provided an opportunity for the geese to expand their range.*

northerly Vesterålen the number of warm days also increased more than fourfold (Figure 15.3). By the end of the 20th century grass growing conditions in Vesterålen, as judged from ambient temperatures, had become comparable to those in the traditional spring range of Helgeland. In this way, a change in climate provided the geese with the opportunity to escape density-dependent processes in the traditional spring staging area.

Comparison of spring habitats

After the shift to new areas began to occur in the 1980–90s, the population was using two contrasting habitats on spring migration, the traditional outer islands and the new agricultural habitats. This resulted in a disparity in the birds' ability to acquire fat and nutrient stores to fuel the last leg of migration and the ensuing breeding season. On average, 17% of the birds managed to breed successfully – which was irrespective of the habitat type

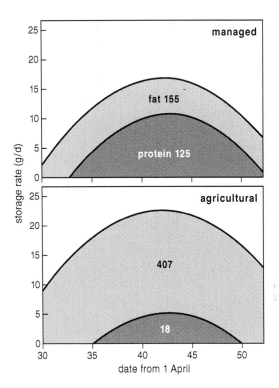

Figure 15.4. *Rate of fat and protein storage for Barnacle Geese staging on managed outer islands and in agricultural habitats in Helgeland, Norway. Accumulated sums (g) for each component are indicated separately. Traditional sites included varying qualities of hay meadows created by Norwegian crofters. Agricultural sites were dominated by well-fertilised pastures. From Prop & Black (1998).*

visited – but it became apparent that the relationship between breeding success and the birds' body condition differed between habitats. Figure 15.4 shows the net accumulation of fat and protein stores for the different habitats. Estimates of net accumulation were obtained from detailed field assessments of foraging rates, diet, chemical composition of foods, and intake rates using the dropping production method (Box 3.3). The accumulation of fat stores was higher on agricultural areas (407g) than on traditional habitats (155g). The amount of protein stores, on the other hand, was higher for birds on the traditional outer islands and very low on agricultural pastures; estimated sums were 125g and 18g, respectively. While Figure 15.5 indicates that the agriculture birds deposited more fat than the outer island birds, Figure 15.6 shows that their extra stores did not result in higher reproduction. The analysis showed a positive relationship between abdominal profile score and reproductive success for outer island birds but not for agriculture birds.[113] Geese utilising agricultural habitat, achieving the highest fatness scores, suffered a depressed probability of successful breeding compared to birds in the traditional habitat, perhaps due to an insufficient protein acquisition.

Evidence from Figure 15.2 suggests that Barnacle Geese in agricultural areas were initially younger, less experienced breeders, which may have depressed the reproductive success for those with the largest abdominal profiles. Perhaps younger birds could not capitalise on

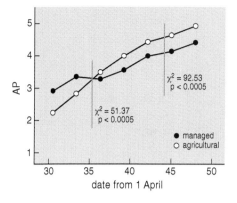

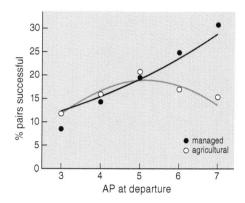

Figure 15.5. *Abdominal profiles of geese on managed outer islands and in agricultural habitats in Helgeland, Norway. Sample sizes were 4,586 and 8,700 geese, respectively. Trends show that whereas agriculture birds arrived with less fat, they departed for the Arctic with larger fat stores than outer island birds. From Prop & Black (1998).*

Figure 15.6. *Probability of successful reproduction in relation to abdominal profile index at time of migration from Helgeland, Norway. Sample sizes for each habitat were: 24, 93, 113, 42, 13, and 40, 55, 109, 133, 116 females, respectively. From Prop & Black (1998).*

the acquisition of large fat stores. Alternatively, an agriculture-derived diet, regardless of age of the bird, may lead to insufficient mechanical (muscle) power because of a deficit of structural proteins. Reduced protein deposition could be due to an unbalanced amino acid composition in the more homogeneous diet found in agricultural areas (Alisauskas & Ankney 1992). In addition, a substantial part of the nitrogen in agricultural crops may be in the form of inorganic compounds – rather than bound to proteins – which impairs the nutritional value (Eichhorn *et al.* 2012). Indeed, others have argued that it is the need for particular nutrients that drives geese to shift from agricultural areas to more natural food sources in spring (McLandress & Raveling 1981, Ebbinge *et al.* 1982, Prins & Ydenberg 1985, McKay *et al.* 1994, Phillips *et al.* 2003). Such a shift by at least some Barnacle Geese toward saltmarsh habitats has also been observed in early spring in Scotland and England prior to migration. Figure 15.7 shows that in April between 50 and 100% of the population move from their earlier use of stubble grains and agricultural pastures to areas of more natural coastal vegetation consisting of a larger variety of grasses, herbs and clover (Owen & Kerbes 1971). Where agricultural land and coastal vegetation are within close range, geese may continue to exploit agricultural grasses throughout the pre-migration period. This is particularly apparent in dry springs when food production drops and geese expand the foraging range to find sufficient food (Prop & Deerenberg 1991). Based on assessments of stable isotopes in food plants and carefully acquired blood samples from 54 female Barnacle Geese captured with cannon nets in habitats of a Wadden Sea island in the Netherlands, Eichhorn *et al.* (2012) showed that each bird had been relying on a mix of agricultural grasses (found on inland pastures) and saltmarsh plants – though to a highly variable extent. It would be worth asking in future what the consequences would be if geese that favoured a predominantly agricultural diet in winter also predominantly made use of agricultural habitats at spring staging sites. It was notable in our study in Helgeland that the 'agriculture' birds, depicted by the trend line in Figure 15.5, arrived at the staging site with a smaller abdominal profile score.

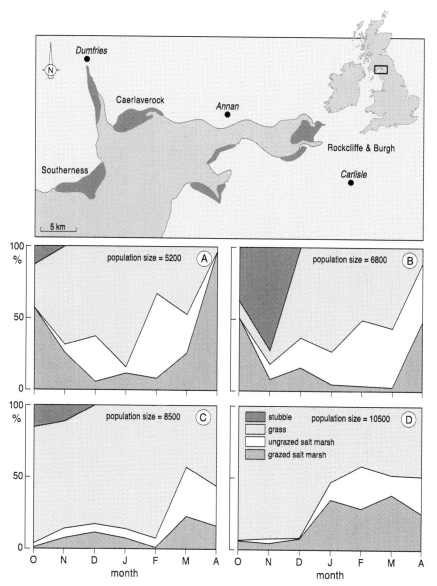

Figure 15.7. *Habitat use by Barnacle Geese, including agricultural crops (post-harvested oat and barley grains), dairy pastures and saltmarsh (merse) habitat. Given is the proportional use of feeding areas in each month in four wintering periods in Scotland and England: (A) 1974–75, (B) 1977–78, (C) 1982–83, (D) 1984–85. Shaded areas consist of grain stubble fields, arable pastures and/or saltmarsh consisting of sea-swept grass-dominated sward. The map indicates key goose habitats on the Solway Firth estuary on the border of Scotland and England. The three main areas where these habitats occur are: (i) Caerlaverock complex (Scotland) including the Caerlaverock National Nature Reserve, the WWT Caerlaverock refuge and nine other farms, totalling over 2,000ha of inland pastures and grain crops and about 1,000ha of managed and unmanaged saltmarsh; (ii) Southerness (Scotland), consisting of approximately 10 farms and the RSPB Mersehead refuge, totalling about 900ha of inland agricultural fields and 400ha of saltmarsh; (iii) Rockcliffe and Burgh (England) totalling 1,300ha of saltmarsh that is managed with livestock grazing. From Owen et al. (1987).*

Other researchers have argued that agricultural habitats may provide geese with more profitable food in comparison with natural habitats (Alisauskas & Ankney 1992, Krapu *et al.* 1995, Robertson & Slack 1995). On the Isle of Islay in Scotland, which has a similar variety of habitat types as in Helgeland, Norway, Fox *et al.* (2005) counted Greenland White-fronted Geese *Anser albifrons flavirostris*, keeping track of grey-plumaged juveniles (four-to eight–month-old goslings) among adults with dark barring on their bellies. By doing so, they were able to determine which flocks produced most young because the flocks on that island faithfully return to particular pastures or coastal strips where the geese acquired a diet of either purposefully sown agricultural grasses and clovers, or a more traditional mix of bog and salt-tolerant plants, respectively. With 10 years of counts they showed that geese wintering exclusively on agricultural habitats produced more young than those making exclusive use of more traditional foods.

Based on our assessment of reproductive success of individually marked Barnacle Geese, we argue that agricultural habitats by themselves do not provide geese with the best possible diet. Birds using agricultural habitats may also be at a disadvantage due to a more expensive energy budget caused by scaring efforts and other disturbances that occur during daily workings on farms (Bos & Stahl 2003, Mini & Black 2009). Madsen (1995, 2001) showed a reduction in fat stores and subsequent reproductive success for Pink-footed Geese utilising farms that do not welcome geese compared to areas where they can feed with fewer interruptions. Observations from Helgeland emphasise the importance of the availability of a variety of food sources enabling the acquisition of a balanced diet for spring staging Barnacle Geese. This points to a clear management objective – to maintain and enhance natural habitats that are adjacent to agricultural areas. It also points to the requirement of finding ways to persuade rural farming communities to support migratory geese. Below, we describe how both of these issues are being addressed in national and international conservation/management initiatives that have been launched on behalf of the Svalbard Barnacle Goose population. We begin by describing a land ethic that incorporates migratory goose populations (*sensu* Leopold 1966).

Wild geese and human perspectives

Wild goose populations have possibly never been as plentiful as they are today. As goose populations continue to rise and their distributions expand, society will ask an increasing number of farmers and ranchers to accommodate geese on their lands. However, if wild geese are in conflict with part of our society then surely their fate becomes less certain. Many managers realise that a population's conservation status is 'unfavourable' when conflicts with humans occur.

In the early days when wild geese were tied to wetlands and coastal areas the few enthusiasts that frequented those areas initially heralded their popular appeal. The once avid waterfowl hunter Sir Peter Scott did much in his later life to convey the majesty of wild geese in his paintings and books (Scott 1938, 1961, Scott & Fisher 1953), and he also achieved it by the establishment of Wildfowl and Wetlands Trust centres near key wintering areas in Britain (Scott 1981). As geese moved closer to urban areas a growing number of people came to view the spectacle and many were struck by the birds' athletic prowess on migration. Since the birds' discovery and utilisation of human-shaped landscapes, our perspective of wild geese has broadened to include both a celebration of 'wilderness values' and a more practical sense of social responsibility. Fox & Madsen (1999) introduce their book, compiling information about European goose populations, with this:

> There can be few people unmoved by the sight of several hundred wild geese lifting into the air in a clamour of cries and thrashing of wings. A wildfowler may be moved by the spectacle and the thrill of the hunt. A birdwatcher may be moved with awe, aware that these birds have bred in far distant arctic regions and travelled many thousands of kilometres to winter quarters. A farmer may be moved by anger, conscious that these geese are devouring part of his livelihood.
>
> In winter, we now know that the Western Palaearctic currently supports [5 million geese from 30] different populations of nine different species [updated from Fox et al. 2010]. In contrast to post-war trends, all but a very few of these populations now show increasing trends. ... Huge flocks of geese descending on crops are of considerable concern to farmers ... An increasingly urbanised general public in Europe, by contrast, place different economical and recreational values on wild geese, generating income for local communities and through government subsidies to agriculture, birdwatching and hunting for example. Whether we like it or not, therefore, geese rank highly on the conservation and agricultural agendas of Europe, and because they are invariably long-distance migrants, they represent a shared natural resource which requires international co-operation in order to resolve conflict. Mechanisms to alleviate agricultural conflict and to manage goose populations and their habitats all require solutions at local, regional and international levels.

If the geese were free to choose, no doubt they would spend most of the winter months at the farms offering the best food and least disturbance (Owen 1971, 1972a,b, 1976a,b). The geese are doing what comes naturally, pressed by tight time and energy schedules – they search for and exploit the best situations on offer. Moser & Kalden (1992), presenting an overview statement at an international symposium about the goose–crop damage conflict, state:

> With no sign of a ceiling to goose populations in Europe, the consequences of inaction are severe; the conflict between farmers and waterfowl will intensify. There is wide agreement that the costs should not be borne by the individual farmer. Wildlife is an integral part of agriculture, but it is still a negligible part of agricultural policies. It is of the utmost importance to broaden the scope of agricultural policies to ensure the conservation of our natural heritage. Farmers need to take account of wildlife as an essential component of the physical and biological environment in which they work. However, the public should contribute significantly to the support of waterfowl to avoid the total cost of this public good falling on individual farmers.

European wildlife managers are successfully developing partnerships among governmental and non-governmental bodies to share responsibility for providing space and food for wild geese in human-mediated landscapes. In Britain, Owen (1977, 1980b, 1990b) consistently supported proposals to safeguard land for wintering geese, either by the creation and management of farm-reserve areas or by making payments to farmers to maintain quality pastures and to tolerate the birds on their land. Managing goose-safe reserves in conjunction with strategic placement of hunting and scaring zones can shift geese into areas that are set aside (Moser & Kalden 1992). A similar call is beginning in North America (Williams *et al.* 1999). Initiatives in the Farm Bill in the USA, including conservation easement programmes, encouraging creation of high-quality wildlife habitat, are particularly heartening (USDA 2014).

Vickery *et al.* (1994) produced a revealing economic analysis of four possible solutions to the problem of growing populations of overwintering geese. They considered the costs to the taxpayer, farmers, conservationists and hunters for four scenarios: maintaining the status quo, culling the geese, paying compensation to farmers, and setting up alternative feeding areas. Their cost–benefit analysis for the British situation at that time showed that the optimal financial solution for the taxpayer was to establish alternative feeding areas (refuges specifically for goose grazing), while the optimal solution for farmers was the payment of compensation (see also Klaassen *et al.* 2008).

The development of goose management in the Netherlands is an interesting case of how a country struggles to meet demands by farmers and conservationists. In this country, the increasing size of Barnacle Goose populations created management challenges that appeared difficult to solve. Within just a couple of decades, winter numbers increased more than tenfold, and a new rapidly growing resident Barnacle Goose population developed. Moreover, claiming damage became a profitable strategy in modern-day society, which placed the government in an uncomfortable position relative to agricultural organisations. A four-year scheme to concentrate all goose species in designated foraging areas failed, as it appeared impossible to accommodate the 1.5 million individuals (Barnacle Geese, White-fronted Geese, Greylag Geese and Pink-footed Geese were involved in the scheme) in an area as small as 80,000ha (van der Jeugd *et al.* 2008). A subsequent plan focused on regulating the resident summer goose populations (Barnacle and Greylag Goose). In the case of the Barnacle Goose the breeding population was to be reduced such that crop damage would not exceed the level of 2011. Among conservationist bodies this plan was hard to accept, but it was traded off against the benefit of the second part of the plan, which was to leave overwintering geese at rest with farmers receiving monetary compensation for damage to their crops plus payment for allowing geese on their land. Although initially adopted by the parties involved, this plan did not get off the ground as farmers' representatives claimed reductions in population size both in summer and winter. We wait to see what will happen with these deliberations.

The conservation action that is occurring on behalf of Barnacle Geese of the Svalbard population includes a mix of the two favoured scenarios. While non-governmental conservation organisations provide alternative feeding areas, governments offer compensation to farmers in the form of incentives to enable geese access to dairy pastures. They are also providing incentive payments to encourage livestock grazers to maintain the attractiveness of marginal saltmarsh habitats, which offer the geese a larger choice of foods and the opportunity to acquire a balanced diet. These habitat management incentive schemes are described in Box 15.1. Through these schemes, managers strive to provide the Svalbard Barnacle Goose population with an adequate amount of habitat that is set aside (i.e. safeguarded) for goose grazing. Two non-governmental conservation organisations were servicing more than half

BOX 15.1 INCENTIVE SCHEMES TO SUPPORT GOOSE FORAGING IN NATURAL AND AGRICULTURAL LANDSCAPES

This section outlines the habitat management agreements between government agencies and private landowners that are currently in place in Scotland, England and Norway, as outlined in the conservation and management action plan for the Svalbard Barnacle Goose population (Black 1998b,c). Conservation, agriculture and countryside agencies are creating ways to support farmers that are willing to accommodate wild geese and other wildlife. Because of these schemes wild geese can be viewed as assets rather than burdens to society. The Scottish and English schemes focus on the north and south side of the Solway Firth, respectively (Figure 15.7), where the geese spend the winter months (September to May). The Norwegian schemes focus on the Helgeland and Vesterålen regions where the birds stop during migration in April and May (Figure 15.3).

Scottish Natural Heritage Barnacle Goose Management Scheme (Scotland) This is a voluntary scheme for farmers working pastures on the Solway Firth, primarily in the vicinity of the WWT Caerlaverock refuge and the Caerlaverock National Nature Reserve. The stated purpose of the scheme is 'to support farming practices that help integrate productive farming with the conservation of Barnacle Geese'. Each field is assigned as a feeding zone, an intermediate zone or a scaring zone based on counts from previous years. Whereas all forms of scaring are allowed in the scaring zone and limited scaring methods in the intermediate zone, no scaring is allowed in the feeding zone. Farmers that participated in the scheme in 1995/96 were paid between £50 and 100 per hectare for fields in intermediate zones, and £160 per hectare for fields in the feeding zones. Scaring equipment was provided by the government agency. The majority of farmers affected by goose grazing entered the scheme and 89% of the relevant area of land was included (880ha) at the inception of the scheme. The total cost of the scheme in 2009/2010 amounted to £241,000, or £35.60 per goose visiting participating farms (Crabtree *et al.* 2010). The scheme has been very successful in concentrating the birds in goose-safe areas. In the first six years, goose densities doubled in the feeding zones, and refuge pastures continued to support more than twice the number of goose-use days compared to all other areas (Cope *et al.* 2003, 2005).

Scottish Natural Heritage Merse (saltmarsh) Management Scheme (Scotland) This scheme is available to farmers and landowners with areas of saltmarsh on the north of the Solway Firth. The purpose of the scheme is to support livestock grazing practices that help integrate productive farming with the special nature conservation interest of the saltmarsh. The scheme is designed to improve this more 'traditional', 'natural' habitat, addressing a range of wildlife interests and, in particular, to increase goose usage by grazing the salt-tolerant grasses down to a height that is more attractive to the geese. The mechanism is to grant annual payments to farmers and landowners that will graze the saltmarsh with livestock to produce a short sward. Participants at the start of the scheme received £30 per hectare. Capital payments were also made toward the costs of necessary fencing and the provision of water supplies for livestock. The majority of farmers (90%) entered the scheme and 75% of the relevant area of land, i.e. 660ha, was included in this scheme.

Countryside Stewardship Scheme (England) The Countryside Commission is the government's statutory adviser on countryside issues, and works to conserve and enhance the beauty of the English countryside and to help people enjoy it. Payments are made for changes to farming and land management practices which produce conservation benefits or improved access to and enjoyment of the countryside. It is open to anyone who can enter a 10-year agreement. The coast is recognised as an important English landscape and one of the scheme's objectives for the coast is to 'manage saltmarshes to sustain their wild, natural appearance, and the plants and animals which they support. Feeding and roosting areas for birds are especially important'. The scheme has been extremely successful on the southern coast of the Solway Firth. It covered 2,698ha of saltmarsh including all of the areas used by Barnacle Geese. Annual payments at the beginning of the scheme were £20 per hectare, with supplementary payments for initial reintroduction of beneficial management (£40 per ha) and for certain capital works. Many of the same goals are achieved under the title *Environmental Stewardship Scheme*, which was launched in 2005.

Outer Island Goose Management Scheme (Norway) The aim of this scheme is to make outer island vegetation more attractive to migrating geese, thus encouraging the birds' traditional use of the area. The County Governor of Nordland initiated their goose management in 1989 in the form of supporting research, initiating a community education programme, financing a herd of weather-hardy sheep, purchasing grass cutting machinery, and paying farmers to move sheep during the goose season. The initiatives were intensified in 1995 by providing incentives for sheep translocations at the Hysvar archipelago, whereby the farmer allows the geese access to three of the main islands that are intensively grazed throughout the year. Prior to the birds' arrival sheep are moved, allowing the birds undisturbed access to the prime foraging sites. Shimmings & Isaksen (2013) reported that only a fraction of the population continues to make use of these traditional island sites.

Herøy/Tenna and Vesterålen Management Scheme (Norway) The aim of this scheme is to redistribute the geese away from prime agricultural fields by coordinated scaring (both passive and active) as well as by improving the attractiveness of once overgrown coastal meadows and pastures in undisturbed goose foraging zones (consisting mainly of *Festuca rubra*). Farmers participating on a voluntary basis since 1996 made the improvements to alternative goose foraging zones by grazing sheep and cattle for most of summer until the onset of harsh weather in late autumn. They also erected scaring devices on the agricultural fields. Initially regional authorities provided funding and materials for goose scaring, cultivation of new foraging areas, fencing and transportation of livestock to alternative sites, and supported efforts to monitor the birds' response to the management scheme. Shimmings (2003) reported that use of the farmers' agricultural fields declined once geese discovered newly established alternative foraging zones. The scheme was expanded in 2006 by offering payments to farmers based on a three-level goose-use of farms. In the first year three million NOK were available, increasing to 3.5 million NOK in the second year (Tombre *et al.* 2013b). Most farmers participated in the scheme, which created goose-safe foraging areas. Farms in Tenna/Herøy that adopted the scheme received more geese than those where prolonged scaring occurred (Shimmings & Isaksen 2013). Tombre *et al.* (2013a,b) reported that the subsidy scheme provided funding that reduced the economic costs caused by the geese.

Largely due to the special relationship that geese have with the marine environment in Norway and the outer island culture and traditions, the Directorate for Nature Management has developed a number of Bird and Animal Protection Areas, Landscape Protection Areas, and a Nature Reserve in Helgeland and other regions. Several of the key sites for Barnacle Geese are included in these areas (Black 1998b,c).

Figure 15.8. *Wintering geese on well-managed grass pasture at WWT Caerlaverock, Scotland (54°N). The adjacent stubble field held leftover barley grains that were also harvested by the geese. Photo points south to distant English hills across the Solway Firth. Photo by Jouke Prop.*

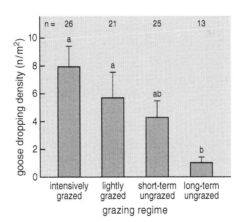

Figure 15.9. *Average goose grazing pressure (+ SE) in 85 transects in relation to livestock grazing regime on saltmarshes in the Wadden Sea area. Grazing pressure was determined by counting goose droppings within plots. Bars that do not share the same letter differed significantly from each other. Goose grazing pressure was positively correlated with the density of preferred grass species. From Bos* et al. *(2005).*

of the winter goose-days on lands that were specifically managed for the geese (Owen *et al.* 1987, Cope *et al.* 2003, 2005). Creating attractive goose habitat has been a goal of the Wildfowl & Wetlands Trust (Figure 15.8) and the Royal Society for the Protection of Birds (RSPB). At these conservation-led farms, cattle and sheep routinely graze refuge pastures and saltmarshes during summer, offering the geese a high-density sward and little dead material during the winter months. Combined with the low fibre content of newly emerging shoots, this provides the geese with a high-quality food supply. Figure 15.9 indicates how geese responded to improvements in pastures and saltmarsh habitats in the Wadden Sea area after experiencing different levels of livestock management (Bos *et al.* 2005). These data show that most goose grazing was attributed to areas with the shorter, well-managed swards.

The suite of schemes, including government subsidies and privately managed refuges, provides an effective solution to reducing goose grazing pressure on individual farms while providing adequate habitat for the geese. In the next section, we describe the overall conservation status of the Svalbard Barnacle Goose population in light of European conservation/management approaches.

Conservation and management for the Svalbard Barnacle Goose population

Waterbird populations in Europe are managed under the auspices of the African-Eurasian Migratory Waterbird Agreement (AEWA) that was created within the Bonn Convention of the United Nations Environment Programme (UNEP) (Boere 2010). Obligations under the Bonn Convention have gained a strong foundation through the Birds and Habitats Directives, which encompass comprehensive legislation to protect wild birds within the European Union. The Agreement requires 'the effective conservation and sustainable use of waterfowl and their habitats' by the almost 100 participating countries. It parallels the North American Waterfowl Management Plan between Canada, USA and Mexico (U.S. Department of the Interior and Environment Canada 1986 and subsequent revisions). While the goal of both of these international schemes is 'conservation of waterfowl populations' there are major differences. The North American scheme includes specific 'upper limit' targets for the size of each population, and bases decisions about harvest and habitat management with these targets in mind. Targets are largely based on long-term averages of the healthy population sizes observed in the 1970s. Given the generally smaller population sizes in the Old World, the African-Eurasian scheme was originally based on a 'lower limit' approach, relying more on the concepts of 'species at risk' and 'minimum viable population size' on which to craft conservation management action. Under pressure from agricultural bodies this policy is changing and the Pink-footed Goose population breeding in Svalbard is the first to have a set population size target (80,000; Madsen & Williams 2012). International action plans are encouraged for populations with unfavourable conservation status – i.e. when a population is less than 25,000 – and for larger populations that are either concentrated on a small number of sites or are in conflict with human interests or other natural resources (e.g. agriculture and fisheries, or impacts on other species). For populations of less than 25,000, the Agreement requires strict protection (no hunting), the identification of a network of suitable habitats that can be safeguarded for the population, identification of remedial measures where human activities (and conflicts) are concerned, emergency response protocols, the support of research and monitoring activities, raising public awareness, and the exchange of information between countries.

The hallmark of African-Eurasian conservation is based on the realisation that waterbird populations require a network of sites throughout their migratory ranges in multiple countries (*sensu* the Ramsar Convention philosophy; Kear *et al.* 2005). Sites are listed as 'Internationally Important' when more than 1% of a waterbird population consistently makes use of them (Scott 1980). Sites designated with this status are most likely to receive protection when conservation measures are taken on behalf of a focal population. Under the criterion outlined in the AEWA, the Svalbard Barnacle Goose population was considered to have unfavourable conservation status for three reasons: (i) fewer than 25,000 individuals in the population, (ii) highly concentrated in small geographical areas throughout the year, and (iii) impact on the livelihood of farmers in rural communities. To comply with AEWA guidelines, in 1995 and 1997 wildlife and land management agencies from Norway, Scotland and England came together to begin drafting a conservation and management action plan for the population. The following long-term objectives were identified (Black 1998b: p. 37):

1. To maintain favourable conservation status for the Svalbard Barnacle Goose population throughout its geographical range.

    The conservation status of the population will be taken as favourable when:

    (i) population parameters indicate that it is maintaining itself on a long-term basis as a viable component of its natural habitat; the population should number at least 25,000 individuals

    (ii) the natural range of the population is neither being reduced nor is likely to be reduced in the foreseeable future, and through habitat management there is potential for reoccupation of formerly utilised sites

    (iii) there are, and will probably continue to be, sufficiently large areas of habitat to maintain the population on a long-term basis.

2. To encourage and support coordinated and collaborative research and monitoring of the population throughout the range states.

3. To raise public awareness of the conservation status of the population throughout the range states.

This plan was prepared because of the realisation that management at one end of the range can affect numbers and distribution at the other end of the range and that it would, therefore, behove all range states to work together, establish common goals and share management strategies. A common approach to reducing conflict with rural agricultural communities was sought. Most notably, management action to reduce conflicts was based on the belief that wider society should pay for the privilege of proximity to wild geese. The plan also focused on safeguarding a network of habitats to reduce the consequences of natural disasters or negative changes in habitat conditions. The goal was to provide some level of protection for areas with most geese. At the time of writing the plan in 1997, 123 goose haunts were identified throughout the birds' range, 62 of which were consistently used by more than 1% of the population and were therefore considered to be of 'international importance'. Twenty-four of the most important sites were being managed specifically for the geese either as national wildlife refuges, sanctuaries, private nature reserves, or private land whose owners were participating in one of the incentive schemes (Black 1998c). There are five more internationally important sites that have been safeguarded by voluntary farmer participation in compensation schemes in the northern haunts of Vesterålen (I. M. Tombre pers. comm.).

An additional 13 important sites were safeguarded in 2002 when Bear Island (Bjørnøya) was declared a Nature Reserve in Norway. In September and October up to 60% of the goose population visited the small areas of vegetation on Bear Island (Owen & Gullestad 1984), which is thought to enable these individuals to survive their long-distance migration (Owen & Black 1989a, 1991a). The habitat used by the geese on Bear Island consists of grass carpets, mainly of *Festuca rubra*, adjacent to seabird cliffs where the vegetation is naturally fertilised (Figure 15.10; Owen & Gullestad 1984). Bear Island is one of Europe's largest seabird colonies with over 2.5 million nesting birds.

In the 17 years since the plan was drafted, the population has continued to grow and expand beyond its traditional boundaries in all parts of its range (Cope *et al.* 2003, 2005, Shimmings & Isaksen 2006, 2013). The population surpassed the 25,000 lower-limit value for the first time in 2002, but declined to 23,900 in 2005 (WWT 2013). The peak count was 35,900 in 2010 and in the autumn of 2012 it was 31,000 (WWT 2013). It was in 2007 when the population and the five-year average remained above the minimum target value, and the AEWA revised the population's status, downgrading it in 2009 from the classification category A2 to B1 to reflect the population now exceeding 25,000 individuals (Crabtree *et al.* 2010). Managers from range states, therefore, have arrived at a new chapter of interpreting AEWA guidelines. No other migratory population in Europe/Africa has made such a remarkable recovery. Now that the population has surpassed the 25,000 value,

Figure 15.10. *Barnacle Geese foraging on the well-fertilised* Festuca rubra *sward found on Bear Island (74°30'N), home to millions of seabirds. The geese use the island vegetation to replenish fat stores during migration. Photo by Brian Morrell.*

statutory agencies have the opportunity to re-evaluate the plan, especially with regard to schemes aimed at reducing conflicts with rural agricultural communities while continuing to safeguard adequate habitats for the Svalbard Barnacle Goose population, as has been accomplished for the Svalbard Pink-footed Goose population (Tombre *et al.* 2013a,b).

Global change, adaptive repertoires and future fortunes

The Barnacle Goose – a species which has existed for 0.5 million years, experiencing a broad range of environments – seems well equipped to adapt to the variation found in current-day breeding and wintering areas. Early Barnacle Geese, pushed by four periods of glaciation, experienced a range of northern and temperate habitats covering much of Europe (Chapter 2). Prior generations of Barnacle Geese would have encountered food plants with a wide range of growth patterns influenced by phases of global warming and cooling. They also coped with a variety of terrestrial and aerial predators that focus on geese as prey. The modern Arctic-breeding Barnacle Goose has a diet of over 51 species of Arctic plants, including seeds, tubers, leaves, forbs and aquatic vegetation (Prop *et al.* 1980). In addition, the newly established temperate-breeding Barnacle Goose populations have learned to exploit a different set of plant species. Thus, the Barnacle Geese of today make use of a broad range of summer habitat types, including expansive stretches of tundra, precarious slopes and cliffs, grass-covered and forested islands in the Baltic, and delta landscapes in the Netherlands. In winter, the sea-swept marshes and a variety of natural and man-made wetlands, meadows, pastures and agricultural crops are preferred.

Ultimately, population growth is enabled by adult survival and the proportion of birds that successfully reproduce, population parameters that are determined by individual decisions made on a daily, annual and lifetime basis (Figure 15.11). We argue that the birds' adaptability to changes in their environment has sustained the growth in our study populations. Depletion of resources at traditional sites may have slowed the rate of growth but the propensity of young birds for sampling and colonising alternative sites and habitats has lessened density-dependent effects. When weather conditions allowed, rates of reproduction were initially higher at the newly discovered breeding sites than at older sites that were close to capacity (Chapter 14). Therefore, rather than the population having a fixed maximum size, individuals' flexibilities enable exploration of new sites and habitats, making the carrying capacity of the population more of a theoretical entity than a certainty (Figure 15.11). Both the large-scale expansion of the spring staging range along the Norwegian mainland coast (Chapter 2) and the explosive spreading of breeding colonies on the borders of the Barents, White, Baltic and North Seas (van der Graaf *et al.* 2006, van der Jeugd *et al.* 2009) are examples of the adaptability of this species.

It is difficult to predict when Barnacle Goose populations will stop growing, due to their willingness to abandon the Arctic lifestyle for temperate areas over 2,000 or more kilometres to the south. Changes in global weather patterns thought to affect plant and animal distributions (Hughes 2000, Shaughnessy *et al.* 2012) will probably continue to shape survival and reproduction in Arctic-breeding geese (Boyd & Madsen 1997, Kéry *et al.* 2006, Sedinger *et al.* 2006). Moreover, high goose densities must have an impact on vulnerable Arctic ecosystems, for example because grazing affects the flow of nutrients (Iacobelli &

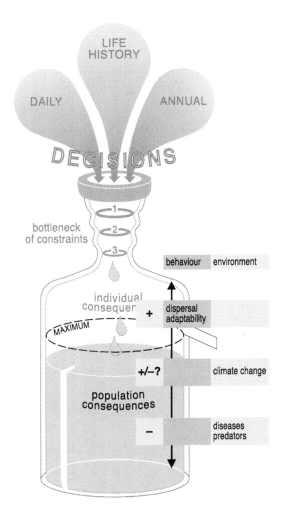

Figure 15.11. *Maximum goose population size in context of the consequences of individual decisions (updated Figure 1.9). The limit to the maximum population size is determined by the carrying capacity of habitats within the normal range. Arrows indicate that the level of the outflow pipe (representing maximum population size) can increase due to the birds' adaptability and dispersal to new habitats and expansion in range, or decrease due to potential limitations from predation pressure and disease. Adaptability enables the geese to break away from limits on reproductive success imposed by density-dependent effects. The effect of climate change is uncertain; it may enhance or limit the viability of populations.*

Jefferies 1991). Any consequent changes in the vegetation might be either positive (Bazely & Jefferies 1986, Person *et al.* 2003) or negative to associated species and to the geese themselves (Kerbes *et al.* 1990, Williams *et al.* 1993). Since reproductive performance is affected by events taking place in more southerly areas (Chapter 11), population growth may also become limited by conditions in winter or, perhaps most likely, along the migratory pathways (Drent *et al.* 2007).

Against the picture of rapidly expanding goose populations, one could almost forget modern threats to their sustainability. In addition to the problem of agricultural conflict with large goose concentrations, the highly gregarious nature of Barnacle Geese in particular makes them vulnerable to disease, pollution and natural catastrophe. Kear *et al.* (2005) review the score of human impacts that potentially threaten the future fortunes of waterfowl populations and call for more research to underpin conservation and management initiatives that will mitigate them. The pursuit of behavioural studies of representative species is crucial to the understanding of processes involved in current evolutionary change (Jefferies & Drent 2006). Insights into individual variation within populations, for example in mate and family relationships, foraging performance, competitive ability and dispersal tendencies, will help us better understand selection that is shaping variance in reproduction and survival. Long-term demographic studies based on marked individuals are needed to generate trends in numbers, recruitment and survival. By considering population parameters in relation to environmental and anthropogenic factors one can identify periods of the life cycle where resources may be limiting. This knowledge enables predictions of population responses to potential conservation and management actions (Pettifor *et al.* 2000, Stillman & Goss-Custard 2010). In addition, well-designed experiments in the field or with captive birds are valuable to test ideas about mechanisms behind individual decisions and population responses. In migratory geese, the interactions between performance on the widely separated wintering, spring and summer areas will make predictions a challenging enterprise.

Summary and conclusions

The Svalbard Barnacle Goose population may be one of the best-protected and most well-managed waterfowl populations in Europe. Its recovery from a low of 300 birds to the current size of over 30,000 is largely due to effective conservation and management action. Many of its favoured haunts are protected or included in conservation schemes. The Barnacle Goose is attributed with a large degree of popular appeal in Britain and Norway. Observers are often awestruck by the bird's athletic prowess in completing 3,100km journeys across Arctic seas. The Barnacle Goose was chosen as the symbol for the regional council of Dumfries and Galloway in Scotland, and in Svalbard the image and lifestyle of the Barnacle Goose is synonymous with wilderness values, which are very much a part of the Norwegian culture. The successful recovery of the population is used as a 'flagship' example for wildlife education and conservation programmes in schools. Many bird watchers and nature lovers view the wild goose spectacle at refuges managed specifically for the geese by two of Britain's leading conservation organisations, although as the population has increased it has overflowed traditional boundaries into neighbouring farms where pastures were managed for dairy cows and sheep rather than geese.

The conservation and management action plan written on behalf of the population emphasises the 'safeguarding' of the birds' favoured habitats and promotes ways of incorporating the geese in the agricultural landscape adjacent to traditional saltmarsh habitats and roosts. Management action to reduce agricultural conflicts is based on the belief that wider society should pay for the privilege of proximity to wild geese. A variety of incentive schemes are in operation to encourage farmers to allow geese on their land, thus creating 'goose-safe' zones adjacent to refuges. The plan emphasises schemes that promote wild geese as assets, rather than burdens, in our society. It stresses the shared responsibility of range states, comprising regional and national government agencies together with private groups and citizens, to ensure a favourable conservation status for this celebrated wild goose population.

Barnacle Geese are well equipped with the behavioural flexibility to discover any new foraging situations that become available, and they have responded favourably to habitat management schemes. However, detailed studies have shown that reliance on sown swards in agricultural pastures resulted in body stores composed mostly of fat and lacking in protein, which limited the benefits of achieving large energy stores. Birds having access to both agricultural land and saltmarsh areas may acquire the nutrients needed for balanced body stores. This information points to the relative ecological benefit of management schemes that aim to enhance the attractiveness of coastal saltmarsh habitats, the original Barnacle Goose habitat. Farmland refuges are a great way to reduce the burden on local farming communities, but management schemes that include both pastures and managed saltmarshes offer a good mix of habitats for the geese.

We set out in this book to describe the attributes of successful individuals in Barnacle Goose society. We have examined how birds vary in terms of mate choices, family associations, diets, foraging performances, competitive abilities, body sizes, predator avoidance, and fidelity to sites, all of which result in a range of possibilities – from birds that never breed during a short lifetime to those that reproduce often during long lifetimes. Indeed, it is the discovery of this breadth of individual differences in behaviours and preferences that has sustained our interest in this model system for so many years. We have learned that the choices that individuals make at each step shape their ability to survive and acquire the resources necessary for completing long-distance migrations and mounting breeding attempts. We show that it is the resourceful nature of wild geese that has continued to fuel population expansions. From the population perspective, it seems that wild geese are now living in a 'land of plenty'. Populations are apparently thriving today because current conditions are within the birds' realm of highly specialised behaviours and body designs. On a population level, it is tempting to conclude that the behavioural repertoire of wild geese will enable populations to cope with an ever-changing world. From an individual perspective, however, the vast majority of geese in these increasing populations strive but fail to reproduce during their lifetimes – only a small minority ever succeed in producing offspring that survive. Even in the best of times, most individuals are hard-pressed to make ends meet. Therefore, if the current conditions (e.g. climate, agricultural preferences and policies) that are accommodating geese change too much then we might end up with a very different verdict about the adaptable nature of wild geese. What is clear overall is that by studying individuals one gains a comprehensive picture of the inner workings of populations.

Statistical analyses

[111] Maximum number of geese counted in northern Norway (Vesterålen) in relation to flyway population size. Regression line for 1998–2013: n = 16, r^2 = 0.63, P < 0.001 (Figure 15.3A). Numbers in Vesterålen from Shimmings & Isaksen (2013) and Tombre *et al.* (2013a).

[112] Probability of a warm day (> 6°C) along the Norwegian coast, between 20 April and 15 May. Regression line: n = 39, r^2 = 0.22, P < 0.05 (Figure 15.3C). The minimum threshold for plant growth is 6°C (Keatinge *et al.* 1979). The slopes of the regressions for the eight selected stations (between 64 and 70°N) were similar ($F_{7,289}$ = 1.03, NS). Extended from Prop *et al.* (1998). Weather data from http://eca.knmi.nl.

[113] Slopes for the fitted logistic regressions for managed outer island and agricultural habitats are 0.236 (χ^2 = 3.63, P = 0.05) and 0.039 (χ^2 = 0.17, NS), respectively (Figure 15.6). From Prop & Black (1998).

Appendix 1

Detailed statistical tests

Source	df	Type III SS	Mean square	F	P
Model	10	209,958	20,996	3.41	0.0002
Error	531	3,272,544	6,163		
Corrected Total	541	3,482,502			
Birth year	8	195,153	24,394	3.96	0.0001
Gosling sex	2	4,069	2,035	0.33	NS

Table 1. *Variation in length of goslings' association (days) with parents in relation to birth year and gosling sex. Length of association was a function of birth year. General Linear Models Procedure (SAS 1996). Class variables included: Birth year (1977; 1986–87; 1989–93; 1995), and gosling sex (Female, Male, Unknown). The Dependent Variable was number of days spent with parents starting from 27 Sept, the first arrival back to Scotland (n = 542 birds, range 0–234 days, R^2 = 0.060, CV = 105.80, root MSE = 78.50 days, average number days with parents 74.20 days). Referred to in Chapter 6, note 42. See Figure 6.6.*

(A) Female goslings

Source	df	Type III SS	Mean square	F	P
Model	12	218,727	18,227	3.28	0.0003
Error	179	993,951	5,553		
Corrected Total	191	1,212,678			
Birth year	6	97,841	16,307	2.94	0.0093
Natal brood size	6	123,166	20,528	3.70	0.0017

(B) Male goslings

Source	df	Type III SS	Mean square	F	P
Model	12	167,658	13,972	2.42	0.0060
Error	200	1,156,215	5,781		
Corrected Total	212	1,323,873			
Birth year	6	40,004	6,667	1.15	NS
Natal brood size	6	97,011	16,169	2.87	0.0124

Table 2. *Variation in offspring–parent length of association for (A) female and (B) male goslings. Length of association was a function of natal brood size, controlled for birth year. General Linear Models Procedure (SAS 1996). Class variables included: birth year (1986; 1989–93; 1995) and brood size (1–6; 8). The Dependent Variable was number of days spent with parents starting from 27 Sept, the first arrival back to Scotland (female and male range 0–231 days; female n = 192, R^2 = 0.180, CV = 79.05, root MSE = 74.52 days, average number days with parents 94.27 days; male n = 213, R^2 = 0.127, CV = 88.79, root MSE = 76.03 days, average number days with parents 85.63 days). Referred to in Chapter 6, note 45. See Figure 6.7.*

Source	df	Type III SS	Mean square	F	P
Model	15	278,804	18,587	3.43	0.0001
Error	324	1,757,433	5,424		
Corrected Total	339	2,036,237			
Birth year	6	127,028	21,171	3.90	0.0009
Natal brood size	3	56,667	18,889	3.48	0.0162
Brood sex ratio	5	74,520	14,904	2.75	0.0190

Table 3. *Variation in goslings–parent length of association in relation to brood size and sex ratio within the brood. General Linear Models Procedure (SAS 1996). A more conservative data set was used in this analysis, including only birds that arrived together with parents, thus excluding the earliest 'orphaned' goslings. Class variables included birth year (1986; 1989–1993; 1995), natal brood size (1–4; 6) and sex ratio of brood (male alone, female alone, male only with other males, female only with other females, male with more than one female, female with more than one male, equal sex ratio). The Dependent Variable was number of days spent with parents starting from 27 Sept, the first arrival back to Scotland (range 0–231 days; n = 340 birds, R^2 = 0.137, CV = 70.87, root MSE 73.65 days, average number of days with parents 103.78 days). Referred to in Chapter 6, note 46.*

Testing length of gosling–parent association in relation to brood sex ratio for each sex separately indicated that the link to association with parents was stronger in males than in females (females n = 165, df = 4, Type III SS = 27,136, Mean Square 6,784, F = 1.15, P = NS; males n = 175, df = 4, Type III SS = 59,945, Mean Square 14,986, F = 2.54, P = 0.0417).

Table 4. *Program MARK model selection testing the viability (survival φ) of 625 young barnacle geese from three months of age to 15 months of age in relation to variables describing early family life with parents. Often, several models in the final set of top-ranked models appear equally plausible, with delta QAICc (quasi-Aikaike's Information Criterion) values near zero and QAICc weights comparable to the best model. The first model indicates that gosling survival did not change during the study (constant survival); however, QAICc weights in the second through seventh models indicate that there was some evidence that survival varied (positively) with brood size, length of association with parents in first year (days) and degree of contact with parents (proportion of observations with parents). This suggests that goslings from larger broods, and those that spent more time with their parents in their first year, were more likely to survive between the ages of three months and 15 months. Referred to in Chapter 6, note 49.*

Model #	Survival	Model parameters	QAICc	Delta QAICc	QAICc Weight	#Par	Deviance	β coefficient	β SE
1	φ	Constant	712.7	0.0	0.26	4	1,084.2	2.035	0.161
2	φ	Brood size	713.2	0.4	0.21	5	1,081.7	0.161	0.129
3	φ	Degree of contact	713.8	1.1	0.15	5	1,082.7	0.124	0.128
4	φ	Length of association	714.5	1.8	0.11	5	1,083.8	0.067	0.129
5	φ	Brood size + Degree of contact	715.1	2.4	0.08	6	1,081.6		
6	φ	Brood size + Length of association	715.2	2.4	0.08	6	1,081.7		
7	φ	Length of association + Degree of contact	715.7	3.0	0.06	6	1,082.5		

Variables were derived from multiple resighting records of goslings; to be included in this analysis goslings were resighted a minimum of five times (and once after 1 Feb) during the goslings' first winter and spring (average 11, maximum 40 resightings). Parameters: φ = apparent survival probability of goslings between arrival in first winter and arrival in second winter in Scotland (age three months to age 15 months). Resighting probability was constant in all models. Terms: 'constant' survival from 1986–1996, brood size effect (one to six goslings in family), length of association (days, ranging from arrival in Scotland already as single gosling to still with parents on departure back to breeding area at age 10 months), and degree of contact with parents (proportion of observations in which gosling was seen with parents; values from 0.1 to 1.0). Model selection was based on c-hat = 1.538. The effects of gosling sex, year (time) and colony (n = five locations) were also tested but dropped during the process of model selection.

Table 5. *Variation in age of first breeding was a function of length of association with parents for (A) males, but not (B) females. General Linear Models Procedure (SAS 1996) for the 1986 cohort. Class variable: natal brood size (0–4). The Dependent Variable was age of first breeding for the 1986 cohort 1988–1995 (range 2–9 yrs; males n = 59, R^2 = 0.254, CV = 138.65, root MSE 2.51 yrs, average age of first breeding = 1.81 yrs; females n = 61, R^2 = 0.046, CV = 160.75, root MSE 2.61 yrs, average age of first breeding = 1.62 yrs). Referred to in Chapter 6, note 50. See Figure 6.8.*

(A) Male goslings

Source	df	Type III SS	Mean square	F	P
Model	5	114	23	3.60	0.0071
Error	53	335	6		
Corrected Total	58	449			
Natal brood size	4	37	9	1.46	NS
Length of association	1	33	33	5.22	0.0264

(B) Female goslings

Source	df	Type III SS	Mean square	F	P
Model	5	18	4	0.53	NS
Error	55	374	7		
Corrected Total	60	392			
Natal brood size	4	18	4	0.66	NS
Length of association	1	1	1	0.18	NS

A similar set of results indicating male age of first breeding was a function of length of association with parents (df = 1, Type III SS = 25, Mean Square 25, F Value 4.91, P = 0.0277) was found with an increased sample (234 birds) from multiple birth-years (1975–77, 1979, 1986–87, 1989–93), while controlling for gosling birth year (df = 10, Type III SS = 305, Mean Square = 30, F Value = 5.96, P < 0.0001). Dependent variable, age of first breeding, was not influenced by the independent variables, (i) lifespan or (ii) original family size.

None of the models for either sex (n = 61 females, n = 59 males) approached the 0.05 level of significance in a step-wise General Linear Model Procedure (SAS 2001). Average lifespan was 7.5 years for non-family goslings (brood size = 0) and 7.6 for family goslings (those still together on arrival in Scotland).

Table 6. *Program MARK results modelling survival (ϕ) of 1972 barnacle goose cohort from Dunøyane colony. The smallest Delta QAICc indicates the best model; in this case the values for a linear and quadratic trend (with bird-age) are almost equal (0 and 0.01, respectively). QAICc weights denote strength of evidence for a given model as the most parsimonious model in the set, and summing QAICc weights of all models containing the variable of interest gives a score of the importance of the variable. For example, the sums of all scores including linear and quadratic trends are 0.56 and 0.41, respectively, whereas models not including an age effect only amount to 0.02. Models including sex were also quite strong, indicating variation between the sexes; sum of QAICc weights with sex (0.58) compared to without sex (0.41). Model selection based on c-hat = 1.085. Sample size: 46 males and 44 females captured as yearlings. Detection probabilities were held constant. Resightings of individually marked geese were recorded on a daily basis throughout the seven-month wintering period and periodically at other times of year. Referred to in Chapter 9, note 77. See Figure 9.1.*

Model	Survival (ϕ)	Delta QAICc	QAICc Weight	#Par	QDeviance
1	ϕ linear trend with age	0.00	0.20	3	332.79
2	ϕ quadratic trend with age	0.01	0.20	4	330.78
3	ϕ sex+linear trend with age	0.20	0.18	4	330.96
4	ϕ sex× linear trend with age	0.26	0.18	5	329.00
5	ϕ sex+quadratic trend with age	0.44	0.16	5	329.18
6	ϕ sex×quadratic trend with age	2.77	0.05	7	327.43
7	ϕ annual variation	5.85	0.01	21	301.33
8	ϕ sex+annual variation	6.46	0.01	22	299.81

Table 7a. *Multi-strata model selection results for female barnacle goose (n = 3,429) cost of reproduction and social feedback loop analysis. Models include parameters for survival (φ), detection (p) and transition (ψ) probabilities for non-breeder (N) and breeder (B) strata. Covariates include time (t, 1975–1993), bird age (a, 3–23 years), annual reproductive success (ars), cumulative reproductive success (crs), and skull length (skull) as an index for body size. Model number 1 is the most parsimonious model for the data, indicating that female survival varied between breeders and non-breeders and that larger females with a larger cumulative reproductive success were more likely to move from a non-breeder to breeder state, and that females with a larger cumulative reproductive success were less likely to move from a breeding to non-breeding status. Likelihood ratio tests (LRT) were performed on competing nested models. P-values < 0.05 indicate that the additional parameter provides significantly more information than the reduced model. P-values > 0.05 indicate no additional information in the additional parameter, so parsimony requires considering the reduced model superior to the model with additional parameter(s). Non-nested models cannot be tested by LRT, and rely on AIC (Aikaike's Information Criterion) to select the model containing the most information. We first examined the possibility that age structure (model number 12) should be used in lieu of time (year) as the main source of temporal variation. Delta AIC value between time and age structured models was 665, meaning time structure explained much more variation in the data than age structure. We continued model selection with time structure in all parameters. Referred to in Chapter 9, note 78. See Figure 9.2.*

Model	Survival (φ)	Resight (p)	Movement (ψ) non→breed	Movement (ψ) breed→non	Delta AICc	AICc Weight	#Par	Deviance	test	χ²	df	P-value
1	φ (t+state)	p N(t) = p B(t)	ψ N to B(t+crs+skull)	ψ B to N (t+crs)	0.0	0.42	84	1,9845.3	1 vs 2	1.03	1	0.31
2	φ (t+state)	p N(t) = p B(t)	ψ N to B(t+crs+skull)	ψ B to N(t+crs+skull)	1.0	0.25	85	1,9844.3				
3	φ (t+state+skull)	p N(t) = p B(t)	ψ N to B(t+crs+skull)	ψ B to N(t+crs)	1.8	0.17	85	1,9845.1	3 vs 4	5.16	2	0.08
4	φ (t+state)	p N(t) = p B(t)	ψ N to B(t+crs)	ψ B to N(t+crs)	2.9	0.10	83	1,9850.3	4 vs 6	69.52	2	0.0001
5	φ (t+state)	p N(t) = p B(t)	ψ N to B(t+crs)	ψ B to N(t+crs+ars)	4.8	0.04	84	1,9850.2	5 vs 4	0.10	1	0.75
6	φ (t+state)	p N(t) = p B(t)	ψ N to B(t)	ψ B to N(t)	68.3	0.00	81	1,9919.8	6 vs 9	6.23	1	0.01
7	φ (t+state+crs)	p N(t) = p B(t)	ψ N to B(t)	ψ B to N(t)	69.7	0.00	82	1,9919.1	7 vs 6	0.68	1	0.41
8	φ (t+state,B+ars)	p N(t) = p B(t)	ψ N to B(t)	ψ B to N(t)	70.3	0.00	82	1,9919.7	8 vs 6	0.05	1	0.82
9	φ N(t)	= p B(t)	ψ N to B(t)	ψ B to N(t)	72.5	0.00	80	1,9926.0	9 vs 10	27.32	40	0.94
10	φ N(t)	= p B(t)	ψ N to B(t)	ψ B to N(t)	88.5	0.00	100	1,9901.2	10 vs 11	2.48	20	1.00
11	φ N(t)	= p B(t)	ψ N to B(t)	ψ B to N(t) TIME	126.9	0.00	120	1,9898.7				
12	φ N(a)	= p B(a)	ψ N to B(a)	ψ B to N(a) AGE	792.0	0.00	120	2,0563.8				
13	φ N(t)	= p B(t)	ψ N to B(t)	ψ B to N(t)	2,394.8	0.00	80	22,248.3	13 vs 10	2,347.13	20	0.0001

Table 7b. *Multi-strata model selection results for male barnacle goose (n = 3,428) cost of reproduction and social feedback loop analysis. Models include parameters for survival (φ), detection (p) and transition (ψ) probabilities for non-breeder (N) and breeder (B) strata. Covariates include time (t, 1975–1993), bird age (a, 3–23 years), annual reproductive success (ars), cumulative reproductive success (crs), and skull length (skull) as an index for body size. Model number 1 is the most parsimonious model, indicating that survival is the same for breeders and non-breeders (see Chapter 14) and that larger males with a larger cumulative reproductive success were more likely to move from a non-breeder to breeder state, and that males with a larger cumulative reproductive success were less likely to move from a breeding to non-breeding status. Likelihood ratio tests (LRT) were performed on competing nested models. P-values < 0.05 indicate that the additional parameter provides significantly more information than the reduced model. P-values > 0.05 indicate no additional information in the additional parameter, so parsimony requires considering the reduced model superior to the model with additional parameter(s). Non-nested models cannot be tested by LRT, and rely on AIC to select the model containing the most information. We first examined the possibility that age structure (model number 12) should be used in lieu of time (year) as the main source of temporal variation. Delta AIC value between time and age structured models was 638, meaning time structure explained much more variation in the data than age structure. We therefore continued model selection with time structure in all parameters. Referred to in Chapter 9, note 78. See Figure 9.2.*

Model	Survival (φ) Non	Survival (φ) breed	Resight (p) non	Resight (p) breed	Movement (ψ) non→breed	Movement (ψ) breed→non	Delta AICc	AICc Weight	#Par	Deviance	test	χ²	df	P-value
1	φ N(t)	= φ B(t)	p N(t)	p B(t)	ψ N to B(t+crs)	ψ B to N(t+crs+ars)	0.0	0.57	103	1,9876.5	1 vs 2	1.15	1	0.28
2	φ N(t)	= φ B(t)	p N(t)	p B(t)	ψ N to B(t+crs+skull)	ψ B to N(t+crs)	0.9	0.37	104	1,9875.4				
3	φ N(t)	= φ B(t+ars)	p N(t)	p B(t)	ψ N to B(t+crs+skull)	ψ B to N(t+crs)	5.6	0.04	106	1,9876.0	3 vs 1	0.55	1	0.46
4	φ N(t+crs)	= φ B(t+ars+crs)	p N(t)	p B(t)	ψ N to B(t+crs+skull)	ψ B to N(t+crs)	7.4	0.01	107	1,9875.7	4 vs 1	0.80	2	0.67
5	φ N(t)	= φ B(t)	p N(t)	p B(t)	ψ N to B(t+crs)	ψ B to N(t+crs+ars)	9.9	0.00	103	1,9886.4	5 vs 6	2.62	1	0.11
6	φ N(t)	= φ B(t)	p N(t)	p B(t)	ψ N to B(t+crs)		10.5	0.00	102	1,9889.1	6 vs 7	69.06	2	0.0001
7	φ N(t)	= φ B(t)	p N(t)	p B(t)	ψ N to B(t)	ψ B to N(t)	75.4	0.00	100	1,9958.1	7 vs 10	16.06	20	0.71
8	φ (t+state)		p N(t)	p B(t)	ψ N to B(t)	ψ B to N(t)	76.1	0.00	101	1,9956.8	9 vs 7	1.36	1	0.24
9	φ N(t+skull)	= φ B(t+skull)	p N(t)	p B(t)	ψ N to B(t)	ψ B to N(t)	77.2	0.00	101	1,9957.9	10 vs 7	0.24	1	0.62
10	φ N(t)	= φ B(t)	p N(t)	p B(t)	ψ N to B(t)	ψ B to N(t)	100.3	0.00	120	1,9942.1	11 vs 10	48.52	20	0.0005
11	φ N(t)	= φ B(t)	p N(t)	p B(t)	ψ N to B(t)	ψ B to N(t)	107.9	0.00	100	1,9990.6				
12	φ N(a)	= φ B(a)	p N(a)	p B(a)	ψ N to B(a)	ψ B to N(a)	745.5	0.00	120	2,0587.2				
13	φ N(t)	= φ B(t)	p N(t)	p B(t)	ψ N to B(t)	ψ B to N(t)	1,730.0	0.00	100	2,1612.7	13 vs 10	1,670.65	20	0.0001

Table 8. *Model selection results for barnacle goose annual movement and fidelity to spring staging sites in Helgeland, Norway (1987–1992). Age differences were included in top model as determined by AIC. The data set was grouped by age class where springtime location (month of May) was for 1) yearlings (22 months of age) in relation to their previous location as 10-month-old goslings, 2) young adults (34 months of age) in relation to their previous location as yearlings, and 3) adults (46+ months of age) in relation to their previous location to their previous location as younger adults. The data set included 2,212 ringed birds resighted over five years at three locations among traditional, outer island and recently discovered agricultural sites. Trends indicated more movement toward the agricultural site, especially by younger birds. Older birds were more site faithful at all locations. See Figure 15.2.*

Model	Survival (φ)	Detectability (p)	Movement (ψ)	AICc	Delta AICc	AICc Weight	#Par	Deviance
1	φ (age+t+site)	p (age+t+site)	ψ (age+site)	5,003.9	0.0	0.52	26	986.1
2	φ (age+t+site)	p (age+t+site)	ψ (age+t+site)	5,004.4	0.6	0.39	30	978.5
3	φ (age+t+site)	p (t+site)	ψ (age+t+site)	5,007.5	3.7	0.08	28	985.7
4	φ (t+site)	p (age+t+site)	ψ (age+t+site)	5,012.4	8.5	0.01	28	990.5
5	φ (age+t+site)	p (age+t+site)	ψ (site)	5,020.3	16.5	0.00	24	1,006.7

Appendix 2

Scientific names of species mentioned in main text. Subspecies names have been included when the subspecies has been mentioned separately.

BIRDS

Aleutian Cackling Goose *Branta hutchinsii leucopareia*
Arctic Skua *Stercorarius parasiticus*
Barnacle Goose *Branta leucopsis*
Black Brant *Branta bernicla nigricans*
Black Guillemot *Cepphus grylle*
Black-capped Chickadee *Poecile atricapilla*
Black-legged Kittiwake *Rissa tridactyla*
Brent Goose *Branta bernicla*
Brunnich's Guillemot *Uria lomvia*
Canada Goose *Branta canadensis*
Cliff Swallow *Hirundo pyrrhonota*
Common Eider *Somateria mollissima*
Common Guillemot *Uria aalge*
Dark-bellied Brent Goose *Branta bernicla bernicla*
Eurasian White-fronted Goose *Anser albifrons albifrons*
Fulmar *Fulmarus glacialis*
Glaucous Gull *Larus hyperboreus*
Golden Eagle *Aquila chrysaetos*
Goldeneye *Bucephala clangula*
Goshawk *Accipiter gentilis*
Great Black-backed Gull *Larus marinus*
Great Skua *Stercorarius skua*
Greater Snow Goose *Chen caerulescens atlantica*
Greater White-fronted Goose *Anser albifrons frontalis*
Greenland White-fronted Goose *Anser albifrons flavirostris*
Greylag Goose *Anser anser*
Gyrfalcon *Falco rusticolus*
Hawaiian Goose *Branta sandvicensis*
Herring Gull *Larus argentatus*
Lesser Black-backed Gull *Larus fuscus*
Lesser Snow Goose *Chen caerulescens caerulescens*
Light-bellied Brent Goose *Branta bernicla hrota*
Little Auk *Alle alle*
Marsh Tit *Parus palustris*
Peregrine Falcon *Falco peregrinus*
Pink-footed Goose *Anser brachyrhynchus*

Puffin *Fratercula arctica*
Red-breasted Goose *Branta ruficollis*
Spanish Imperial Eagle *Aquila heliaca*
Steller's Jay *Cyanocitta stelleri*
Tundra Bean Goose *Anser fabalis rossicus*
Western Screech Owl *Otus kennicottii*
White-fronted Goose *Anser albifrons*
White-tailed Eagle *Haliaeetus albicilla*
Wigeon *Anas penelope*
Wood Duck *Aix sponsa*

MAMMALS

Arctic (or Polar) Fox *Alopex lagopus*
Polar Bear *Ursus maritimus*
Raccoon Dog *Nyctereutes procyonoides*
Red Fox *Vulpes vulpes*
Reindeer *Rangifer tarandus platyrhynchus*

PLANTS

Alpine Bistort *Bistorta vivipara*
Arctic Bluegrass *Poa arctica*
Arctic Mouse-ear *Cerastium arcticum*
Arctic Saltmarsh Sedge *Carex subspathacea*
Buttercup *Ranunculus* spp.
Calliergon spp.
Clover *Trifolium* spp.
Creeping Alkali Grass *Puccinellia phryganodes*
Foxtail *Alopecurus borealis*
Mouse-ear *Cerastium* spp.
Poa spp.
Polar Scurvygrass *Cochlearia groenlandica*
Polar Willow *Salix polaris*
Puccinellia maritima
Purple Saxifrage *Saxifraga oppositifolia*
Red Fescue *Festuca rubra*
Ryegrass *Lolium perenne*
Saltgrass *Puccinellia* spp.
Sea Plantain *Plantago maritima*
Timothy Grass *Phleum pratense*
Tufted Saxifrage *Saxifraga cespitosa*
Tundra Grass *Dupontia pelligera*
Tundra grass *Dupontia* spp.
Variegated Horsetail *Equisetum variegatum*
White Clover *Trifolium repens*

References

Adams, D. C. & Anthony, C. D. 1996. Using randomization techniques to analyse behavioural data. *Animal Behaviour* 51: 733–738.

Adolph, E. F. 1947. Urges to eat and drink in rats. *American Journal of Physiology* 157: 110–121.

Akesson, T. R. & Raveling, D. G. 1982. Behaviors associated with seasonal reproduction and long-term monogamy in Canada geese. *Condor* 84: 188–196.

Alisauskas, R. T. & Ankney, C. D. 1990. Body size and fecundity in lesser snow geese. *Auk* 107: 440–443.

Alisauskas, R. T. & Ankney, C. D. 1992. Spring habitat use and diets of midcontinent adult lesser snow geese. *Journal of Wildlife Management* 56: 43–54.

Anderholm, S., Marshall, R. C., van der Jeugd, H. P., Waldeck, P., Larsson, K. & Andersson, M. 2009a. Nest parasitism in the barnacle goose: evidence from protein fingerprinting and microsatellites. *Animal behaviour* 78: 167–174.

Anderholm, S., Waldeck, P., van der Jeugd, H. P., Marshall, R. C., Larsson, K. & Andersson, M. 2009b. Colony kin structure and host-parasite relatedness in the barnacle goose. *Molecular Ecology* 18: 4955–4963.

Anderson, B. G. 1981. Late Weichselian ice sheets in Eurasia and Greenland. In: Denton, G. H. & Hughes, T. J. (eds) *The Last Great Ice Sheets*. Wiley, New York, pp. 1–65.

Andersson, M. 1984. Brood parasitism within species. In: Barnard, C. J. (ed) *Producers and Scroungers: Strategies for exploitation and parasitism*. Croom Helm, London, pp. 195–228.

Andersson, M. 1994. *Sexual Selection*. Princeton University Press, Princeton.

Andersson, M. 2001. Relatedness and the evolution of conspecific brood parasitism. *American Naturalist* 158: 599–614.

Andersson, M. & Åhlund, M. 2000. Host–parasite relatedness shown by protein fingerprinting in a brood parasitic bird. *Proceedings of the National Academy of Sciences* 97: 13188–13193.

Andersson, M. & Åhlund, M. 2001. Protein fingerprinting: a new technique reveals extensive conspecific brood parasitism. *Ecology* 82: 1433–1442.

Ankney, C. D. 1977. The use of nutrient reserves by breeding male lesser snow geese *Chen caerulescens caerulescens*. *Canadian Journal of Zoology* 55: 1984–1987.

Ankney, C. D. & MacInnes, D. C. 1978. Nutrient reserves and reproductive performance of female lesser snow geese. *Auk* 95: 459–471.

Avital, E., Jablonka, E. & Lachman, M. 1998. Adopting adoption. *Animal Behaviour* 55: 1451–1459.

Bang, C., Gullestad, N., Larsen, T. & Norderhaug, M. 1963. Norsk Ornitologisk Spitsbergen ekspedisjon, sommeren 1962. *Norsk Polarinstitutt Årbok 1962*: 93–119.

Barry, T. W. 1962. Effect of late seasons on Atlantic brant reproduction. *Journal of Wildlife Management* 26: 19–26.

Bazely, D. R. & Jefferies, R. L. 1986. Changes in the composition and standing crop of salt-marsh communities in response to the removal of a grazer. *Journal of Ecology* 74: 693–706.

Beauchamp, G. 1997. Determinants of intraspecific brood amalgamation in waterfowl. *Auk* 114: 11–21.

Bell, M. C., Fox, A. D., Owen, M., Black, J. M. & Walsh, A. J. 1993. Approaches to estimation of survival in two arctic-nesting goose species. In: Lebreton, J-D. & North, P. M. (eds) *Marked Individuals in the Study of Bird Population Dynamics*. Birkhauser Verlag, Basel, pp. 141–155.

Belthoff, J. R. & Dufty, A. M. Jr. 1998. Corticosterone, body condition and locomotor activity: a model for dispersal in screech owls. *Animal Behaviour* 55: 405–415.

Bennike, O., Bjorck, S., Bocher, J., Hansen, L., Heinemeier, J. & Wohlfahrt, B. 1999.

Early Holocene plant and animal remains from North-east Greenland. *Journal of Biogeography* 26: 667–677.

Bêty, J., Gauthier, G. & Giroux, J. F. 2003. Body condition, migration, and timing of reproduction in snow geese: A test of the condition-dependent model of optimal clutch size. *American Naturalist* 162: 110–121.

Bigot, E., Hausberger, M. & Black, J. M. 1995. Exuberant youth: the example of triumph ceremonies in barnacle geese. *Ethology, Ecology and Evolution* 7: 11–25.

Bize, P., Roulin, A., Tella, J. L., Bersier, L-F. & Richner, H. 2004. Additive effects of ectoparasites over reproductive attempts in the long-lived alpine swift. *Journal of Animal Ecology* 73: 1080–1088.

Black, J. M. 1988. Preflight signalling in swans: a mechanism for group cohesion and flock formation. *Ethology* 79: 143–157.

Black, J. M. 1996. Introduction: pair bonds and partnerships. In: Black, J. M. (ed) *Partnerships in Birds. The Study of Monogamy.* Oxford University Press, Oxford, pp. 3–20.

Black, J. M. 1998a. Movement of barnacle geese between colonies in Svalbard and the colonization process. *Norsk Polarinstitutt Skrifter* 200: 115–128.

Black, J. M. 1998b. Flyway plan for the Svalbard population of barnacle geese. A summary. *Norsk Polarinstitutt Skrifter* 200: 29–40.

Black, J. M. 1998c. *Conservation and Management Action Plan for the Svalbard Population of Barnacle Geese.* Prepared for the Directorate for Nature Management and Scottish Natural Heritage.

Black, J. M. 2001. Fitness consequences of long-term pair bonds in barnacle geese: monogamy in the extreme. *Behavioral Ecology* 12: 640–645.

Black, J. M. & Barrow J. H. Jr. 1985. Visual signalling in Canada geese for the coordination of family units. *Wildfowl* 36: 35–41.

Black, J. M., Carbone, C., Owen, M. & Wells, R. 1992. Foraging dynamics in goose flocks: the cost of living on the edge. *Animal Behaviour* 44: 41–50.

Black, J. M., Choudhury, S. & Owen, M. 1996. Do geese benefit from life-long monogamy? In: Black, J. M. (ed) *Partnerships in Birds. The Study of Monogamy.* Oxford University Press, Oxford, pp. 91–117.

Black, J. M., Cooch, E. G., Loonen, M. J. J. E., Drent, R. H. & Owen, M. 1998. Body size variation in barnacle goose colonies: Evidence for local saturation of habitats. *Norsk Polarinstitutt Skrifter* 200: 129–140.

Black, J. M., Deerenberg, C. & Owen, M. 1991. Foraging behaviour and site selection of barnacle geese in a traditional and newly colonized spring staging area. *Ardea* 79: 349–358.

Black, J. M. & Owen, M. 1984. The importance of the family unit to barnacle goose offspring, a progress report. *Norsk Polarinstitutt Skrifter* 181: 79–84.

Black, J. M. & Owen, M. 1987. Determinant factors of social rank in goose flocks, acquisition of social rank in young geese. *Behaviour* 102: 129–146.

Black, J. M. & Owen, M. 1988. Variations in pair bond and agonistic behaviors in barnacle geese on the wintering grounds. In: Weller, M. (ed) *Waterfowl in Winter.* University of Minnesota Press, Minneapolis, pp. 39–57.

Black, J. M. & Owen, M. 1989a. Parent–offspring relationships in wintering barnacle geese. *Animal Behaviour* 37: 187–198.

Black, J. M. & Owen, M. 1989b. Agonistic behaviour in goose flocks: assessment, investment and reproductive success. *Animal Behaviour* 37: 199–209.

Black, J. M. & Owen, M. 1995. Reproductive performance and assortative pairing in relation to age in barnacle geese. *Journal of Animal Ecology* 64: 234–244.

Black, J. M., Patterson, D., Shimmings, P. & Rees, E. C. 1999. Barnacle goose numbers on the Solway Firth: 1990–1996. *Scottish Birds* 20: 63–72.

Bliss, L. C. 1981. North American and Scandinavian tundras and polar deserts. In: Bliss, L. C., Heal, O. W. & Moore, J. J. (eds) *Tundra Ecosystems: a comparative analysis.* Cambridge University Press, Cambridge, pp. 8–24.

Boere, G. C. 2010. *The History of the Agreement on the Conservation of African-Eurasian Migratory Waterbirds*. UNEP / AEWA Secretariat, Bonn, Germany.

Bonsall, M. B., French, D. R. & Hassell, M. P. 2002. Metapopulation structures affect persistence of predator–prey interactions. *Journal of Animal Ecology* 71: 1075–1084.

Bos, D., Loonen, M. J. J. E., Stock, M., Hofeditz, F., van der Graaf, A. J. & Bakker, J. P. 2005. Utilisation of Wadden Sea salt marshes by geese in relation to livestock grazing. *Journal for Nature Conservation* 13: 1–15.

Bos, D. & Stahl, J. 2003. Creating new foraging opportunities for dark-bellied brent *Branta bernicla* and barnacle geese *Branta leucopsis* in spring: insights from a large-scale experiment. *Ardea* 91: 153–165.

Boyd, H. 1953. On encounters between wild white-fronted geese in winter flocks. *Behaviour* 5: 85–29.

Boyd, H. 1964. Barnacle geese caught in Dumfriesshire in February, 1963. *Wildfowl* 15: 76–76.

Boyd, H. & Madsen, J. 1997. Impacts of global change on Arctic breeding bird populations and migration. In: Oechel, W. C., Callaghan, T., Gilmanov, T., Holten, J. I., Maxwell, B., Molau, U. & Sveinbjornsson, B. (eds) *Global Change and Arctic Terrestrial Ecosystems*. Springer-Verlag, Berlin, pp. 201–217.

Bried, J., Pontier, D. & Jouventin, P. 2003. Mate fidelity in monogamous birds: a re-examination of the Procellariiformes. *Animal Behaviour* 65: 235–246.

British Trust for Ornithology (BTO) 2013. Summary for all ringing recoveries for barnacle goose *Branta leucopsis*. http://blx1.bto.org/ring/countyrec/resultsall/rec1670all.htm Accessed January 2014.

Brown, C. R. & Brown, M. B. 1996. *Coloniality in the Cliff Swallow. The Effect of Group Size on Social Behavior*. Chicago University Press, Chicago.

Brownie, C., Hines, J. E., Nichols, J. D., Pollock, K. H. & Hestbeck, J. B. 1993. Capture–recapture studies for multiple strata including non-Markovian transitions. *Biometrics* 49: 1173–1187.

Buchsbaum, R., Wilson, J. & Valiela, I. 1986. Digestibility of plant constituents by Canada geese and Atlantic brant. *Ecology* 67: 386–393.

Burley, N. 1977. Parental investment, mate choice and mate quality. Proceedings of the *National Academy of Sciences* 74: 3476–3479.

Bustnes, J. O., Persen, E. & Bangjord, G. 1995. Results from the survey of the light-bellied brent goose and barnacle goose populations on Tusenøyane and South-western Svalbard in July 1995. *NINA Opdragsmelding* 378: 1–13.

Butler, P. J. & Woakes, A. J. 1998. Behaviour and energetics of Svalbard barnacle geese during their autumn migration. *Norsk Polarinstitutt Skrifter* 200: 165–174.

Cabot, D. 1984. *Biological Expedition to Jameson Land, Greenland 1984*. Barnacle Books, Dublin.

Carbone, C., Thompson, W. A., Zadorina, L. Rowcliffe, J. M. 2003. Competition, predation risk and patterns of flock expansion in barnacle geese. *Journal of Zoology* 259: 301–308.

Caro, T. 1999. *Behavioral Ecology and Conservation Biology*. Oxford University Press, Oxford.

Cézilly, F., Dubois, F. & Pagel, M. 2000. Is mate fidelity related to site fidelity? A comparative analysis in Ciconiiforms. *Animal Behaviour* 59: 1143–1152.

Chapman, D. F., Clark, D. A., Land, C. A. & Dymock, N. 1983. Leaf and tiller growth of *Lolium perenne* and *Agrostis* spp. and leaf appearance rates of *Trifolium repens* in set-stocked and rotationally grazed hill pastures. *New Zealand Journal of Agricultural Research* 26: 159–168.

Choudhury, S. 1995. Divorce in birds: a review of hypotheses. *Animal Behaviour* 50: 431–429.

Choudhury, S. & Black, J. M. 1993. Mate choice strategies in geese: evidence for a "partner-hold" strategy. *Animal Behaviour* 46: 747–757.

Choudhury, S. & Black, J. M. 1994. Barnacle geese preferentially pair with familiar associates from early life. *Animal Behaviour* 48: 81–88.

Choudhury, S., Black, J. M. & Owen, M. 1992. Do barnacle geese pair assortatively? Lessons from a long-term study. *Animal Behaviour* 44: 171–173.

Choudhury, S., Black, J. M. & Owen, M. 1996. Body size, compatibility and fitness in barnacle geese. *Ibis* 138: 700–709.

Choudhury, S., Jones, C., Black, J. M. & Prop, J. 1993. Adoption of young and intraspecific nest parasitism in barnacle geese. *Condor* 95: 860–868.

Clausen, K. K. & Clausen, P. 2013. Earlier Arctic springs cause phenological mismatch in long-distance migrants. *Oecologia* 173: 1101–1112.

Clausen, P. & Percival, S. M. 1998. Changes in distribution and habitat use of Svalbard light-bellied brent geese *Branta bernicla hrota*, 1980–1995: Driven by *Zostera* availability? *Norsk Polarinstitutt Skrifter* 200: 253–276.

Clemmons, J. R. & Buchholz, R. 1997. *Behavioral Approaches to Conservation in the Wild*. Cambridge University Press, Cambridge.

Clutton-Brock, T. H. 1988. *Reproductive Success*. University of Chicago Press, Chicago.

Collias, N. E. & Jahn, L. R. 1959. Social behavior and breeding success in Canada geese (*Branta canadensis*) confined under semi-natural conditions. *Auk* 76: 478–509.

Cooch, E. G., Lank, D. B. & Cooke, F. 1996. Intraseasonal variation in the development of sexual size dimorphism in a precocial bird – evidence from the lesser snow goose. *Journal of Animal Ecology* 65: 439–450.

Cooch, E. G., Lank, D. B., Dzubin, A., Rockwell, R. F. & Cooke, F. 1991a. Body size variation in lesser snow geese: environmental plasticity in gosling growth rates. *Ecology* 72: 503–512.

Cooch, E. G., Lank, D. B., Rockwell, R. F. & Cooke, F. 1991b. Long-term decline in body size in a snow goose population: evidence for environmental degradation? *Journal of Animal Ecology* 60: 483–496.

Cooch, E. G., Lank, D. B., Rockwell, R. F. & Cooke, F. 1992. Is there a positive relationship between body size and fecundity in lesser snow geese? *Auk* 109: 667–673.

Cooke, F., Davies, J. C. & Rockwell, R. F. 1990. Response to Alisauskas and Ankney. *Auk* 107: 444–446.

Cooke, F., Findlay, C. S., Rockwell, R. F. & Abraham, K. F. 1983. Life history studies of the lesser snow goose (*Anser caerulescens caerulescens*). II. Colony structure. *Behavioral Ecology and Sociobiology* 12: 153–159.

Cooke, F., Rockwell, R. F. & Lank, D. 1995. *The Snow Geese of La Pérouse Bay*. Oxford University Press, Oxford.

Cope, D. R., Pettifor, R. A., Griffin, L. R. & Rowcliffe, J. M. 2003. Integrating farming and wildlife conservation: the barnacle goose management scheme. *Biological Conservation* 110: 113–122.

Cope, D., Vickery, J. & Rowcliffe, R. 2005. From conflict to coexistence: a case study of geese and agriculture in Scotland. In: Woodroffe, M., Thirgood, S. & Rabinowitz, A. (eds) *People and Wildlife, Conflict or Coexistence?* Cambridge University Press, Cambridge.

Cote, J., Clobert, J., Brodin, T., Fogarty, S. & Sih, A. 2010. Personality-dependent dispersal: characterization, ontogeny and consequences for spatially structured populations. *Philosophical Transactions of the Royal Society B* 365: 4065–4076.

Coulson, J. C. 1966. The influence of the pair-bond and age on the breeding biology of the kittiwake gull, *Risa tridactyla*. *Journal of Animal Ecology* 35: 269–279.

Coulson, J. C. 1972. The significance of the pair bond in the kittiwake gull. In: Iwasa, Y., Pomiankowski, A. & Nee, S. (eds) *Proceedings of the XVth International Ornithological Congress*. Brill, Leiden, pp. 424–433.

Cowan, P. J. 1973. Parental calls and the approach behavior of young Canada geese: a laboratory study. *Canadian Journal of Zoology* 51: 647–650.

Crabtree, B., Humphreys, L., Moxey, A. & Wernham, C. 2010. *2010 Review of Goose Management Policy in Scotland*. Report to Scottish government. British Trust for Ornithology, Stirling, Scotland.

Crawley, M. J. 1983. *Herbivory: the dynamics of animal-plant interactions.* Blackwell Scientific Publications, Oxford.

Daan, S., Dijkstra, C. & Tinbergen, J. M. 1990. Family planning in the kestrel (*Falco tinnunculus*): The ultimate control of covariation of laying date and clutch size. *Behaviour* 114: 83–116.

Dalhaug, L., Tombre, I. T. & Erikstad, K. E. 1996. Seasonal decline in clutch size of the barnacle goose in Svalbard. *Condor* 98: 42–47.

Davies, N. B. 2000. *Cuckoos, Cowbirds and Other Cheats.* T & AD Poyser, London.

Davies, N. B., Krebs, J. R. & West, S. A. 2012. *Introduction to Behavioural Ecology.* 4th edn. Blackwell Scientific Publications, Oxford.

de Jongh, A. 1983. Mechanism of grazing in three species of geese (*Anser albifrons, Branta bernicla, Branta leucopsis*) (in Dutch). MSc Thesis, University of Groningen, Groningen.

de Kogel, C. H. 1997. Long-term effects of brood size manipulation on morphological development and sex-specific mortality of offspring. *Journal of Animal Ecology* 66: 167–178.

de Veer, G. 1598. *Waerachtighe beschrijvinghe van drie seylagien, ter werelt noyt soo vreemt ghehoort.* Reprinted in 1997, Amsterdam.

Deerenberg, C., Apanius, V., Daan, S. & Bos, N. 1997. Reproductive effort decreases antibody responsiveness. *Proceedings of the Royal Society B, Biological Sciences* 264: 1021–1029.

Delany, S. & Scott, D. A. 2006. *Waterfowl Population Estimates.* 4th edn. Wetlands International, Wageningen, The Netherlands.

Denton, G. H. & Hughes, T. J. 1981. The Arctic ice sheet: an outrageous hypothesis. In: Denton, G. H. & Hughes, T. J. (eds) *The Last Great Ice Sheets.* Wiley, New York, pp. 437–467.

Dickey, M. H., Gauthier, G. & Cadieux, M. C. 2008. Climatic effects on the breeding phenology and reproductive success of an arctic-nesting goose species. *Global Change Biology* 14: 1973–1985.

Dingemanse, N. J., Both, C., Drent, P. J. & Tinbergen, J. M. 2004. Fitness consequences of avian personalities in a fluctuating environment. *Proceedings of the Royal Society B, Biological Sciences* 271: 847–852.

Dingemanse, N. J. & Réale, D. 2005. Natural selection and animal personality. *Behaviour* 142: 1159–1184.

Dormann, C. F. 2003. Consequences of manipulations in carbon and nitrogen supply for concentration of anti-herbivore defence compounds in *Salix polaris. Ecoscience* 10: 312–318.

Drent, R. H., Black, J. M., Prop, J. & Loonen, M. J. J. E. 1998. Barnacle geese *Branta leucopsis* on Nordenskiöldkysten, west Spitsbergen – in thirty years from colonisation to saturation. *Norsk Polarinstitutt Skrifter* 200: 105–114.

Drent, R. H. & Daan, S. 1980. The prudent parent: energetic adjustments in avian breeding. *Ardea* 68: 225–252.

Drent, R. H., Erichhorn, G., Flagstad, A., Van der Graaf, A. J., Litvin, K. E. & Stahl, J. 2007. Migratory connectivity in Arctic geese: spring stopovers are the weak links in meeting targets for breeding. *Journal of Ornithology* 148: S501–S514.

Drent, R. H. & Prins, H. H. T. 1987. The herbivore as prisoner of its food supply. In: van Andel, J., Bakker, J. & Snaydon, R. W. (eds) *Disturbance in Grasslands; Species and Population Responses.* Dr. W. Junk Publishing Company, Dordrecht, pp. 133–149.

Drent, R. H. & Prop, J. 2008. Barnacle goose *Branta leucopsis* survey on Nordenskiöldkysten, west Spitsbergen 1975–2007: breeding in relation to carrying capacity and predator impact. *Circumpolar Studies* 4: 59–83.

Drent, R. H. & Swierstra, P. 1977. Goose flocks and food finding: field experiments with barnacle geese in winter. *Wildfowl* 28: 15–20.

Drent, R. H., Weijand, B. & Ebbinge, B. 1978. Balancing the energy budget of arctic breeding geese throughout the annual cycle: a progress report. *Verhandlungen der Ornithologischen Gesellschaft in Bayern* 23: 239–264.

Duckworth, R. A. 2009. Maternal effects and range expansion: a key factor in a dynamic process? *Philosophical Transactions of the Royal Society B* 364: 1075–1086.

Dufty, A. M. Jr. & Belthoff, J. R. 2001. Proximate mechanisms of natal dispersal: the role of body condition and hormones. In: Clobert, J., Danchin, E., Dhondt, A. A. & Nichols, J. D. (eds) *Dispersal*. Oxford University Press, Oxford, pp. 217–229.

Durant, D., Fritz, H., Blais, S. & Duncan, P. 2003. The functional response in three species of herbivorous Anatidae: effects of sward height, body mass and bill size. *Journal of Animal Ecology* 72: 220–231.

Dzubin, A. & Cooch, E. G. 1992. *Measurements of Geese. General Field Methods*. California Waterfowl Association, Sacramento.

Eadie, J. M., Kehoe, F. P. & Nudds, T. D. 1988. Pre-hatch and post-hatch brood amalgamation in North American Anatidae: a review of hypotheses. *Canadian Journal of Zoology* 66: 1709–1721.

Ebbinge, B. S. 1985. Factors determining the population size of arctic-breeding geese, wintering in western Europe. *Ardea* 73: 121–128.

Ebbinge, B. S. 1991. The impact of hunting on mortality rates and spatial distribution of geese wintering in the western Palearctic. *Ardea* 79: 197–210.

Ebbinge, B. S. 1992. Regulation of numbers of dark-bellied brent geese *Branta bernicla bernicla* on spring staging areas. *Ardea* 80: 203–228.

Ebbinge, B. S., Canters, K. & Drent, R. 1975. Foraging routines and estimated daily food intake in barnacle geese wintering in the northern Netherlands. *Wildfowl* 26: 5–19.

Ebbinge, B. S., St Joseph, A., Prokosch, P. & Spaans, B. 1982. The importance of spring staging areas for arctic-breeding geese, wintering in western Europe. *Aquila* 89: 249–258.

Ebbinge, B. S., van Biezen, J. B. & van der Voet, H. 1991. Estimation of annual adult survival rates of barnacle geese *Branta leucopsis* using multiple resightings of marked individuals. *Ardea* 79: 73–112.

Eichhorn, G., Drent, R. H., Stahl, J., Leito, A. & Alerstam, T. 2009. Skipping the Baltic: the emergence of a dichotomy of alternative spring migration strategies in Russian barnacle geese. *Journal of Animal Ecology* 78: 63–72.

Eichhorn, G., Meijer, H. A. J., Oosterbeek, K. & Klaassen, M. 2012. Does agricultural food provide a good alternative to a natural diet for body store deposition in geese? *Ecosphere* 3: art35. http://dx.doi.org/10.1890/ES11-00316.1 Accessed January 2014.

Elverhøi, A., Dowdeswel, J. A., Funder, S., Mangerud, J. & Stein, R. 1998. Glacial and oceanic history of the Polar North Atlantic margins: an overview. *Quaternary Science Review* 17: 1–10.

Ely, C. R. 1989. Extra-pair copulation in the greater white-fronted goose. *Condor* 91: 990–991.

Ens, B. J., Choudhury, S. & Black, J. M. 1996. Mate fidelity and divorce in birds. In: Black, J. M. (ed) *Partnerships in Birds. The Study of Monogamy*. Oxford University Press, Oxford, pp. 344–401.

Ens, B. J., Safriel, U. N. & Harris, M. P. 1993. Divorce in the long-lived and monogamous oystercatcher, *Haematopus ostralegus*: incompatibility or choosing the better option? *Animal Behaviour* 45: 1199–1217.

Falconer, D. S. & Mackay, T. F. C. 1996. *Introduction to Quantitative Genetics*. 4th edn. Longman, New York.

Feige, N., van der Jeugd, H. P., van der Graaf, A.J., Larsson, K., Leito, A. & Stahl, J. 2008. Newly established breeding sites of the Barnacle Goose *Branta leucopsis* in North-western Europe – an overview of breeding habitats and colony development. *Vogelwelt* 129: 244–252.

Féret, M., Bêty, J., Gauthier, G., Giroux, J. F. & Picard, G. 2005. Are abdominal profile useful to assess body condition of spring staging greater snow geese? *Condor* 107: 694–702.

Ferrer, M. 1993. Ontogeny of dispersal distances in young Spanish imperial eagles. *Behavioral Ecology and Sociobiology* 32: 259–263.

Festa-Bianchet, M. & Apollonio, M. 2003. *Animal Behavior and Wildlife Conservation*. Island Press, Covello.

Finney, G. & Cooke, F. 1978. Reproductive habits in the snow goose: the influence of female age. *Condor* 80: 147–158.

Fischer, H. 1965. Das triumphgeschrei der graugans (*Anser anser*). *Zeitschrift für Tierpsychologie* 22: 247–304.

Fondell, T. F., Flint, P. L., Sedinger, J. S., Nicolai, C. A. & Schamber, J. L. 2011. Intercolony variation in growth of black brant goslings on the Yukon-Kuskokwim Delta, Alaska. *Journal of Wildlife Management* 75: 101–108.

Forslund, P. 1993. Vigilance in relation to brood size and predator abundance in the barnacle goose *Branta leucopsis*. *Animal Behaviour* 45: 965–973.

Forslund, P. & Larsson, K. 1991. The effect of mate change and new partner's age on reproductive success in the barnacle goose, *Branta leucopsis*. *Behavioral Ecology* 2: 116–122.

Forslund, P. & Larsson, K. 1995. Intraspecific nest parasitism in the barnacle goose: behavioural tactics of parasites and hosts. *Animal Behaviour* 50: 509–517.

Fox, A. D. 1993. Pre-nesting feeding selectivity of pink-footed geese *Anser brachyrhynchus* in artificial grasslands. *Ibis* 135: 417–423.

Fox, A. D. & Bergersen, E. 2005. Lack of competition between barnacle geese *Branta leucopsis* and pink-footed geese *Anser brachyrhynchus* during the pre-breeding period in Svalbard. *Journal of Avian Biology* 36: 173–178.

Fox, A. D., Boyd, H. & Bromley, R. G. 1995. Mutual benefits of associations between breeding and non-breeding white-fronted geese *Anser albifrons*. *Ibis* 137: 151–156.

Fox, A. D., Ebbinge, B. S., Mitchell, C., Heinicke, T., Aarvak, T., Colhoun, K., Clausen, P., Dereliev, S., Faragó, S., Koffijberg, K., Kruckenberg, H., Loonen, M. J. J. E., Madsen, J., Mooij, J., Musil, P., Nilsson, L., Pihl, S. & van der Jeugd, H. 2010. Current estimates of goose population sizes in western Europe, a gap analysis and an assessment of trends. *Ornis Svecica* 20: 115–127.

Fox, A. D. & Madsen, J. 1999. Introduction. In: Madsen, J., Cracknell, G. & Fox, A. D. (eds) *Goose Populations of the Western Palearctic*. Wetlands International, Wageningen, The Netherlands; and National Environmental Research Institute, Rönde, Denmark, pp. 8–18.

Fox, A. D., Madsen, J., Boyd, H., Kuijken, E., Norriss, D. W., Tombre, I. M. & Stroud, D. A. 2005. Effects of agricultural change on abundance, fitness components and distribution of two arctic-nesting goose populations. *Global Change Biology* 11: 881–893.

Francis, C. M., Richards, M. H., Cooke, F. & Rockwell, R. F. 1992. Long-term changes in survival rates of lesser snow geese. *Ecology* 73: 1346–1362.

Frigerio, D. B., Weiß, B. M. & Kotrschal, K. 2001. Spatial proximity among adult siblings in greylag geese (*Anser anser*): Evidence for female bonding? *Acta Ethologica* 3: 121–125.

Gabriel, P. O. & Black, J. M. 2012. Behavioural syndromes, partner compatibility, and reproductive performance in Steller's jays. *Ethology* 118: 76–86.

Ganter, B., Boyd, W. S., Baranyuk, V. V. & Cooke, F. 2005. First pairing in snow geese *Anser caerulescens*: at what age and at what time of year does it occur? *Ibis* 147: 57–66.

Ganter, B., Larsson, K., Syroechkovsky, E. V., Litvin, K. E., Leito, A. & Madsen, J. 1999. Barnacle goose *Branta leucopsis*: Russia/Baltic. In: Madsen, J., Cracknell, G. & Fox, A. D. (eds) *Goose Populations of the Western Palearctic* (Wetlands International Publication 48). Wetlands International, Wageningen, The Netherlands; and National Environmental Research Institute, Rönde, Denmark, pp. 270–283.

García-Navas, V. & Sans, J. J. 2011. Females call the shots: breeding dispersal and divorce in blue tits. *Behavioral Ecology* 22: 932–939.

Gauthier, G. & Bédard, J. 1990. The role of phenolic compounds and nutrients in determining food preference in greater snow geese. *Oecologia* 84: 553–558.

Gauthier, G., Bêty, J. & Hobson, K. 2003. Are greater snow geese capital breeders? New evidence from a stable-isotope model. *Ecology* 84: 3250–3264.

Gauthier, G., Pradel, R., Menu, S. & Lebleton, J. D. 2001. Seasonal survival of greater snow geese and effect of hunting under dependence in sighting probability. *Ecology* 82: 3105–3119.

Gill, J. A., Norris, K., Potts, P. M., Gunnarsson, T. G., Atkinson, P. W. & Sutherland, W. J. 2001. The buffer effect and large-scale population regulation in migratory birds. *Nature* 412: 436–438.

Gill, M., Beever, D. E. & Osbourn, D. F. 1989. The feeding value of grass and grass products. In: Holmes, W. (ed) *Grass: its production and utilization*. Blackwell Scientific Publications, Oxford, pp. 89–129.

Goosemap 2013. Site-specific information for geese occurring on Svalbard. http://goosemap.nina.no Accessed January 2014.

Gosling, L. M. & Sutherland, W. J. 2000. *Behaviour and Conservation*. Cambridge University Press, Cambridge.

Goss-Custard, J. D. 1996. *The Oystercatcher. From Individuals to Populations*. Oxford University Press, Oxford.

Goss-Custard, J. D. & Sutherland, W. J. 1997. Individual behaviour, populations and conservation. In: Krebs, J. R. & Davies, N. B. (eds) *Behavioural Ecology: An evolutionary approach*. Blackwell Scientific Publications, Oxford, pp. 373–395.

Gowaty, P. 1996. Battles of the sexes and origins of monogamy. In: Black, J. M. (ed) *Partnerships in Birds. The Study of Monogamy*. Oxford University Press, Oxford, pp. 21–52.

Greenwood, P. J. 1980. Mating systems, philopatry and dispersal in birds and mammals. *Animal Behaviour* 28: 1140–1162.

Greenwood, P. J. 1987. Inbreeding, philopatry and optimal outbreeding in birds. In: Cooke, F. & Buckley, P. A. (eds) *Avian Genetics: a population and ecological approach*. Academic Press, London, pp. 207–222.

Gregoire, P. E. & Ankney, C. D. 1990. Agonistic behavior and dominance relationships among lesser snow geese during winter and spring migration. *Auk* 107: 550–560.

Griffin, L. R. 2008. Identifying the pre-breeding areas of the Svalbard barnacle goose *Branta leucopsis* between mainland Norway and Svalbard: an application of GPS satellite-tracking techniques. *Vogelwelt* 129: 226–232.

Griffin, L. R. 2009. Svalbard barnacle goose monitoring 2008/09. *WWT Goose News* 8: 14.

Griffin, L. R. 2012. Svalbard barnacle goose monitoring 2011/12. *WWT Goose News* 11: 22–23.

Griffin, L. R. 2013. Svalbard barnacle goose monitoring 2012/13. *WWT Goose News* 12: 19–20.

Griffith, S. C., Owens, J. P. F. & Thuman, K. A. 2002. Extra pair paternity in birds: a review of interspecific variation and adaptive function. *Molecular Ecology* 11: 2195–2212.

Griggs, K. M. 2003. Differential allocation of parental care in western Canada geese. MS Thesis, Humboldt State University, Arcata.

Gullestad, N., Owen, M. & Nugent, M. 1984. Numbers and distribution of barnacle geese *Branta leucopsis* on Norwegian staging islands and the importance of the staging area to the Svalbard population. *Norsk Polarinstitutt Skrifter* 181: 57–65.

Gustafsson, L., Nordling, D., Andersson, M. S., Sheldon, B. C. & Qvarnström, A. 1994. Infectious diseases, reproductive effort and the cost of reproduction in birds. *Philosophical Transactions of the Royal Society B* 346: 323–331.

Hahn, S., Loonen, M. J. J. E. & Klaassen, M. 2011. The reliance on distant resources for egg formation in high Arctic breeding barnacle geese *Branta leucopsis*. *Journal of Avian Biology* 42: 159–168.

Halliday, T. R. 1983. The study of mate choice. In: Bateson, P. (ed) *Mate Choice*. Cambridge University Press, Cambridge, pp. 3–32.

Halliday, T. R. & Slater, P. J. B. 1983. *Animal Behaviour. Volume 3: Genes, Development and Learning*. Blackwell Scientific Publications, Oxford.

Hamilton, W. D. 1971. Geometry for the selfish herd. *Journal of Theoretical Biology* 31: 295–311.

Hammond, S. 1990. An investigation of intra-flock behaviour of barnacle geese. MSc Thesis, University of Reading.

Hanson, H. C. 1967. Characters of age, sex, and sexual maturity in Canada geese. *Illinois Natural History Survey Biological Notes* 49: 1–15.

Hanssen, S. A., Folstad, I. & Erikstad, K. E. 2003. Reduced immunocompetence and cost of reproduction in common eiders. *Oecologia* 136: 457–464.

Harrison, J. M. 1974. *Conservation and Wildfowling in Action*. WAGBI Conservation Publication.

Hatten, L. & Norderhaug, A. 2001. The islands of Vega – a coastal cultural landscape in decay or a treasure in present-day society? (In Norwegian) Utmark 2001–1, http://www.utmark.org (November 2006).

Hausberger, M. & Black, J. M. 1990. Do females turn males on and off in barnacle goose social display? *Ethology* 84: 232–238.

Hausberger, M., Black, J. M. & Pichard, J. P. 1991. Bill opening and sound spectrum in barnacle goose loud calls: individuals with 'wide mouths' have higher pitched voices. *Animal Behaviour* 42: 319–322.

Hausberger, M., Richard, J. P., Black, J. M. & Quirs, R. 1994. Quantitative analysis of individuality in barnacle goose loud calls. *Bioacoustics* 5: 247–260.

Heg, D., Bruinzeel, L. W. & Ens, B. J. 2003. Fitness consequences of divorce in the oystercatcher, *Haematopus ostralegus*. *Animal Behaviour* 66: 175–184.

Hestbeck, J. B. & Malecki, R. A. 1989. Estimated survival rates of Canada geese within the Atlantic flyway. *Journal of Wildlife Management* 53: 91–96.

Hewitt, G. M. & Butlin, R. K. 1997. Causes and consequences of population structure. In: Krebs, J. R. & Davies, N. B. (eds) *Behavioural Ecology: An evolutionary approach*. 4th edn. Blackwell Scientific Publications, Oxford, pp. 350–372.

Higgins, K. F., Oldemeyer, J. L., Jenkins, K. J., Clambey, G. K. & Harlow, R. F. 1996. Vegetation sampling and measurement. In: Bookhout, T. A. (ed) *Research Management Techniques for Wildlife and Habitats*. The Wildlife Society, Bethesda, pp. 567–591.

Hirschenhauser, K., Mostl, E. & Kotrschal, K. 1999. Within-pair testosterone covariation and reproductive output in greylag geese *Anser anser*. *Ibis* 141: 577–586.

Hocking, B. 1968. Insect-flower associations in the high Arctic with special reference to nectar. *Oikos* 19: 359–388.

Holling, C. S. 1959. Some characteristics of simple types of predation and parasitism. *Canadian Entomologist* 91: 385–398.

Holm, T. E. & Madsen, J. 2013. Incidence of embedded shotgun pellets and inferred hunting kill amongst Russian/Baltic barnacle geese *Branta leucopsis*. *European Journal of Wildlife Research* 59: 77–80.

Horn, H. S. 1984. Some theories about dispersal. In: Swingland, I. R. & Greenwood, P. J. (eds) *The Ecology of Animal Movement*. Clarendon Press, Oxford, pp. 54–62.

Hübner, C. E. 2006. The importance of pre-breeding sites in the arctic barnacle goose *Branta leucopsis*. *Ardea* 94: 701–713.

Hübner, C. E., Tombre, I. M., Griffin, L. R., Loonen, M. J. J. E., Shimmings, P. & Jónsdóttir, I. S. 2010. The connectivity of spring stopover sites for geese heading to arctic breeding grounds. *Ardea* 98: 145–154.

Hughes, L. 2000. Biological significance of global warming: is the signal already apparent? *Trends in Ecology and Evolution* 15: 56–61.

Iacobelli, A. & Jefferies, R. L. 1991. Inverse salinity gradients in coastal marshes and the death of stands of *Salix*: the effects of grubbing by geese. *Journal of Ecology* 79: 61–73.

Iason, G. R., Manso, T., Sim, D. A. & Hartley, F. G. 2002. The functional response does not predict the local distribution of European rabbits (*Oryctolagus cuniculus*) on grass swards: Experimental evidence. *Functional Ecology* 16: 394–402.

Ims, R. A. & Hjermann, D. Ø. 2001. Condition-dependent dispersal. In: Clobert, J., Danchin, E., Dhondt, A. A. & Nichols, J. D. (eds) *Dispersal*. Oxford University Press, Oxford, pp. 201–216.

Inglis, I. R. 1977. The breeding behaviour of the pink-footed goose: behavioural correlates of nesting success. *Animal Behaviour* 25: 747–764.

Inglis, I. R. & Lazarus, J. 1981. Vigilance and flock size in brent geese: the edge effect. *Zeitschrift Tierpsychologie* 57: 193–200.

Iverson, S. A., Gilchrist, H. G., Smith, P. A., Gaston, A. J. & Forbes, M. R. 2014. Longer ice-free seasons increase the risk of nest depredation by polar bears for colonial breeding birds in the Canadian Arctic. *Proceedings of the Royal Society B, Biological Sciences* 281.

Janetos, A. C. 1980. Strategies of female mate choice: a theoretical analysis. *Behavioral Ecology and Sociobiology* 7: 107–112.

Jefferies, R. L. & Drent, R. H. 2006. Arctic geese, migratory connectivity and agricultural change: calling the sorcerer's apprentice to order. *Ardea* 94: 537–554.

Johnson, J. C. & Raveling, D. G. 1988. Weak family associations in cackling geese during winter: effects of body size and food resources on goose social organization. In: Weller, M. (ed) *Waterfowl in Winter*. University of Minnesota Press, Minneapolis, pp. 71–89.

Johnstone, R. 1997. The evolution of animal signals. In: Krebs, J. R. & Davies, N. B. (eds) *Behavioural Ecology: An evolutionary approach*. 4th edn. Blackwell Scientific Publications, Oxford, pp. 155–178.

Jonker, R. M., Kraus, R. H. S., Zhang, Q., van Hooft, P., Larsson, K., van der Jeugd, H. P., Kurvers, R. H. J. M., van Wieren, S. E., Loonen, M. J. J. E., Crooijmans, R. P. M. A., Ydenberg, R. C., Groenen, M. A. M. & Prins, H. H. T. 2013. Genetic consequences of breaking migratory traditions in barnacle geese *Branta leucopsis*. Molecular Ecology 22: 5835–5847.

Jonker, R. M., Kuiper, M. W., Snijders, L., Van Wieren, S. E., Ydenberg, R. C. & Prins, H. H. T. 2011. Divergence in timing of parental care and migration in Barnacle geese. *Behavioral Ecology* 22: 326–331.

Jourdain, F. C. R. 1922. The breeding habits of the barnacle goose. *Auk* 34: 166–171.

Kalmbach, E. 2006. Why do goose parents adopt unrelated goslings? A review of hypotheses and empirical evidence, and new research questions. *Ibis* 148: 66–78.

Kear, J. 1965. Internal food reserves of hatching mallard ducklings. *Journal of Wildlife Management* 29: 523–528.

Kear, J. 1990. *Wildfowl and Man*. T & AD Poyser, Calton.

Kear, J., Jones, T. & Matthews, G. V. T. 2005. Conservation and management. In: Kear, J. (ed) *Ducks, Geese and Swans*. Oxford University Press, Oxford, pp. 152–171.

Keatinge, J. D. H., Stewart, R. H. & Garrett, M. K. 1979. The influence of temperature and soil water potential on the leaf extension rate of perennial ryegrass in Northern Ireland. *Journal of Agricultural Science* 92: 175–183.

Kerbes, R. H., Kotanen, P. M. & Jefferies, R. L. 1990. Destruction of wetland habitats by lesser snow geese: a keystone species on the west coast of Hudson Bay. *Journal of Applied Ecology* 27: 242–258.

Kéry, M., Madsen, J. & Lebreton, J. D. 2006. Survival of Svalbard pink-footed geese *Anser brachyrhynchus* in relation to winter climate, density and land-use. *Journal of Animal Ecology* 75: 1172–1181.

Kirby, J. S., Stattersfield, A. J., Butchart, S. H. M., Evans, M. I., Grimmett, R. F. A., Jones, V. R., O'Sullivan, J., Tucker, G. M. & Newton, I. 2008. Key conservation issues for migratory land- and waterbird species on the world's migration flyways. *Bird Conservation International* 18: S49–S73.

Klaassen, M., Bauer, S., Madsen, J. & Possingham, H. 2008. Optimal management of a goose flyway: migrant management at minimum cost. *Journal of Applied Ecology* 45: 1446–1452.

Komdeur, J. & Hatchwell, B. J. 1999. Kin recognition: function and mechanism in avian societies. *Trends in Ecology and Evolution* 14: 237–241.

Kotrschal, K., Hirschenhauser, K. & Mostl, E. 1998. The relationship between social stress and dominance is seasonal in greylag geese. *Animal Behaviour* 55: 171–176.

Kraaijeveld, K. & Mulder, R. A. 2002. The function of triumph ceremonies in the black swan. *Behaviour* 139: 45–54.

Kralj-Fišer, S., Scheiber, I. B. R., Blejec, A., Moestl, E. & Kotrschal, K. 2007. Individualities in a flock of free-roaming greylag geese: behavioral and physiological consistency over time and across situations. *Hormones and Behavior* 51: 239–248.

Kralj-Fišer, S., Weiß, B. M. & Kotrschal, K. 2010. Behavioural and physiological correlates of personality in greylag geese (*Anser anser*). *Journal of Ethology* 28: 363–370.

Krapu, G. L., Reinecke, K. J., Jorde, D. G. & Simpson, S. G. 1995. Spring-staging ecology of mid-continent greater white-fronted geese. *Journal of Wildlife Management* 59: 736–746.

Kriengwatana, B., Wada, H., Macmillan, A. & MacDougall-Shackleton, S. A. 2013. Juvenile nutritional stress affects growth rate, adult organ mass, and innate immune function in zebra finches (*Taeniopygia guttata*). *Physiological and Biochemical Zoology* 86: 769–781.

Kristiansen, J. N., Fox, A. D. & Nachman, G. 2000. Does size matter? Maximising nutrient and biomass intake by shoot selection amongst herbivorous geese. *Ardea* 88: 119–125.

Kristiansen, J. N., Walsh, A. J., Fox, A. D., Boyd, H. & Stroud, D. A. 1999. Variation in the belly barring of the Greenland white-fronted goose *Anser albifrons flavirostris*. *Wildfowl* 50: 21–28.

Kumari, E. 1971. Passage of the barnacle goose through the Baltic area. *Wildfowl* 22: 35–43.

Kurvers, R. H. J. M., Adamczyk, V. M. A. P., van Wieren, S. E. & Prins, H. H. T. 2011. The effect of boldness on decision-making in barnacle geese is group-size-dependent. *Proceedings of the Royal Society B, Biological Sciences* 278: 2018–2024.

Kurvers, R. H. J. M., Nolet, B. A., Prins, H. H. T., Ydenberg, R. C. & van Oers, K. 2012. Boldness affects foraging decisions in barnacle geese: an experimental approach. *Behavioral Ecology* 23: 1155–1161.

Kurvers, R. H. J. M., Adamczyk, V. M. A. P., Kraus, R. H. S., Hoffman, J. I., van Wieren, S. E., van der Jeugd, H. P., Amos, W., Prins, H. H. T. & Jonker, R. M. 2013. Contrasting context dependence of familiarity and kinship in animal social networks. *Animal Behaviour* 86: 993–1001.

Lack, D. 1954. *The Natural Regulation of Animal Numbers*. Oxford University Press, Oxford.

Lamprecht, J. 1984. Measuring the strength of social bonds: experiments with hand-reared goslings (*Anser indicus*). *Behaviour* 91: 115–127.

Lamprecht, J. 1986a. Structure and causation of the dominance hierarchy in a flock of bar-headed geese (*Anser indicus*). *Behaviour* 96: 28–48.

Lamprecht, J. 1986b. Social dominance and reproductive success in a goose flock (*Anser indicus*). *Behaviour* 97: 50–65.

Lamprecht, J. 1989. Mate guarding in geese: awaiting female receptivity, protection of paternity or support of female feeding? In: Rasa, A. E., Vogel, C. & Voland, E. (eds) *The Sociobiology of Sexual and Reproductive Strategies*. Chapman and Hall, London, pp. 48–66.

Lamprecht, J. 1990. Predicting current reproductive success of goose pairs *Anser indicus* from male and female reproductive history. *Ethology* 85: 123–131.

Landvik, J. Y., Bondevik, S., Elverhøi, A., Fjeldskaar, W., Mangerud, J., Salvigsen, O., Siegert, M. J., Svendsen, J. I. & Vorren, T. 1998. The last glacial maximum of Svalbard and the Barents Sea area: ice sheet extent and configuration. *Quaternary Science Review* 17: 43–75.

Lang, A. & Black, J. M. 2001. Foraging efficiency in barnacle geese: a functional response to sward height and an analysis of sources of individual variation. *Wildfowl* 52: 7–20.

Lank, D. B., Bousfield, M. J., Cooke, F. & Rockwell, R. F. 1991. Why do snow geese adopt eggs? *Behavioral Ecology* 2: 181–187.

Lank, D. B., Cooch, E. G., Rockwell, R. F. & Cooke, F. 1989b. Environmental and demographic correlates of intraspecific nest parasitism in lesser snow geese *Chen caerulescens caerulescens*. *Journal of Animal Ecology* 58: 29–45.

Lank, D. B., Mineau, P., Rockwell, R. F. & Cooke, F. 1989a. Intraspecific nest parasitism and extra-pair copulation in lesser snow geese. *Animal Behaviour* 37: 74–89.

Larsson, K. 1993. Inheritance of body size in the barnacle goose under different environmental conditions. *Journal of Evolutionary Biology* 6: 195–208.

Larsson, K. 1996. Genetic and environmental effects on timing of wing moult in the barnacle goose. *Heredity* 76: 100–107.

Larsson, K. & Forslund, P. 1991. Environmentally induced morphological variation in the barnacle goose, *Branta leucopsis*. *Journal of Evolutionary Biology* 4: 619–636.

Larsson, K. & Forslund, P. 1992. Genetic and social inheritance of body size and egg size in the barnacle goose (*Branta leucopsis*). *Evolution* 46: 235–244.

Larsson, K. & Forslund, P. 1994. Population dynamics of the barnacle goose *Branta leucopsis* in the Baltic area: density-dependent effects on reproduction. *Journal of Animal Ecology* 63: 954–962.

Larsson, K., Forslund, P., Gustafsson, L. & Ebbinge, B. S. 1988. From the high Arctic to the Baltic: the successful establishment of a barnacle goose *Branta leucopsis* population on Gotland, Sweden. *Ornis Scandinavica* 19: 182–189.

Larsson, K., Tegelstöm, H. & Forslund, P. 1995. Intraspecific nest parasitism and adoption of young in the barnacle goose: effects on survival and reproductive performance. *Animal Behaviour* 50: 1349–1360.

Larsson, K. & van der Jeugd, H. P. 1998. Continuing growth of the Baltic barnacle goose population: number of individuals and reproductive success in different colonies. *Norsk Polarinstitutt Skrifter* 200: 213–219.

Larsson, K., van der Jeugd, H. P., van der Veen, I. T. & Forslund, P. 1998. Body size declines despite positive directional selection on heritable size traits in a barnacle goose population. *Evolution* 52: 1169–1184.

Lazarus, J. & Inglis, I. R. 1978. The breeding behaviour of the pink-footed goose:

parental care and vigilant behaviour during the fledging period. *Behaviour* 65: 62–88.

Lebreton, J. D., Hines, J. E., Pradel, R., Nichols, J. D. & Spendelow, J. A. 2003. Estimation by capture-recapture of recruitment and dispersal over several sites. *Oikos* 101: 253–264.

Lehikoinen, A., Christensen, T. K., Öst, M., Kilpi, M., Saurola, P. & Vattulainen, A. 2008. Large-scale change in the sex ratio of a declining eider *Somateria mollissima* population. *Wildlife Biology* 14: 288–301.

Leito, A. 1996. *The Barnacle Goose in Estonia*. Estonia Maritima. Publication of the West-Estonian Archipelago Biosphere Reserve.

Leito, A. 2011. *Monitoring of Geese in Estonia, Report 2011*. Tartu, 15 pp. (in Estonian with English summary) http://eelis.ic.envir.ee/seireveeb/aruanded/13151_Hanede_seire_l6pparuanne_2011.doc Accessed January 2014.

Leopold, A. 1966. *A Sand County Almanac: with essays on conservation from Round River*. Oxford University Press, Oxford.

Lepage, D., Desrochers, A. & Gauthier, G. 1999. Seasonal decline of growth and fledging success in snow geese *Anser caerulescens*: An effect of date or parental quality? *Journal of Avian Biology* 30: 72–78.

Lepage, D., Gauthier, G. & Menu, S. 2000. Reproductive consequences of egg-laying decisions in snow geese. *Journal of Animal Ecology* 69: 414–427.

Lepage, D., Gauthier, G. & Reed, A. 1998. Seasonal variation in growth of greater snow goose goslings: The role of food supply. *Oecologia* 114: 226–235.

Lessells, C. M. 1985. Natal and breeding dispersal of Canada geese *Branta canadensis*. *Ibis* 127: 31–41.

Lessells, C. M. 1986. Brood size in Canada geese: a manipulation experiment. *Journal of Animal Ecology* 55: 669–689.

Lessells, C. M. 1987. Parental investment, brood size and time budgets: behaviour of lesser snow goose families. *Ardea* 75: 189–203.

Levins, R. 1969. Some demographic and genetic consequences of environmental heterogeneity for biological control. *Bulletin of the Entomological Society of America* 15: 237–240.

Lie, R. W. 1989. Animal remains from the post-glacial warm period in Norway. *Fauna Norvegica Series A* 10: 45–56.

Lindberg, M. S., Sedinger, J. S., Derksen, D. V. & Rockwell, R. F. 1998. Natal and breeding philopatry in a black brant (*Branta bernicla nigricans*) metapopulation. *Ecology* 79: 1839–1904.

Lindholm, A., Gauthier, G. & Desrochers, A. 1994. Effects of hatch date and food supply on gosling growth in Arctic-nesting greater snow geese. *Condor* 96: 898–908.

Lings, G. H. 1935. Lapland, Bear Island and Spitsbergen. In: Steward, E. S. (ed) *Bird Life of the Far North*. Peregrine Books, Leeds.

Loonen, M. J. J. E. 2005. Arctic geese: quick meals for predators: Geese, the green wave and the price of parenthood. In: Drent, R., Bakker, J. P., Piersma, T. & Tinbergen, J. M. (eds) *Seeking Nature's Limits: Ecologists in the field*. KNNV Publishing, Utrecht, pp. 73–78.

Loonen, M. J. J. E., Bruinzeel, L. W., Black, J. M. & Drent, R. H. 1999. The benefit of large broods in barnacle geese: a study of natural and experimental manipulations. *Journal of Animal Ecology* 68: 753–768.

Loonen, M. J. J. E., Larsson, K., van der Veen, I. T. & Forslund, P. 1997a. Timing of wing moult and growth of young in Arctic and temperate breeding barnacle geese. In: Loonen, M. J. J. E. (ed) Goose breeding ecology: overcoming successive hurdles to raise goslings. PhD Thesis, University of Groningen, Groningen, pp. 137–153.

Loonen, M. J. J. E., Oosterbeek, K. & Drent, R. H. 1997b. Variation in growth and adult size: evidence for density dependence. *Ardea* 85: 177–192.

Loonen, M. J. J. E., Tombre, I. M. & Mehlum, F. 1998. Population development of an arctic barnacle goose colony: the interaction between density and predation. *Norsk Polarinstitutt Skrifter* 200: 67–80.

Lorenz, K. 1966. *On Aggression*. Harcourt Publishers, New York.

Lorenz, K. & Tinbergen, N. 1938/1970. Taxis and instinctive behavior pattern in egg-rolling by the greylag goose. In: Lorenz, K. (ed) *Studies in Animal and Human Behavior, Vol. 1*. Harvard University Press, Cambridge, pp. 316–359.

Løvenskiold, H. L. 1964. Avifauna Svalbardensis. *Norsk Polarinstitutt Skrifter* 129: 1–460.

Lyon, B. E. & Eadie, J. M. 2008. Conspecific brood parasitism in birds: a life-history perspective. *Annual Review of Ecology, Evolution, and Systematics* 39: 343–363.

MacArthur, R. H. 1972. *Geographical Ecology: Patterns in the distribution of species*. Harper & Row, New York.

Madsen, J. 1995. Impacts of disturbance on migratory waterfowl. *Ibis* 133: S67–S74.

Madsen, J. 2001. Spring migration strategies in pink-footed geese *Anser brachyrhynchus* and consequences for spring fattening and fecundity. *Ardea* 89: 43–55.

Madsen, J., Bregnballe, T. & Mehlum, F. 1989. Study of the breeding ecology and behaviour of the Svalbard population of light-bellied brent goose *Branta bernicla hrota*. *Polar Research* 7: 1–21.

Madsen, J., Cracknell, G. & Fox, A. D. 1999. *Goose Populations of the Western Palearctic. A Review of Status and Distribution* (Wetlands International Publication 48). Wetlands International, Wageningen, The Netherlands; and National Environmental Research Institute, Kalö, Denmark.

Madsen, J., Frederiksen, M. & Ganter, B. 2002. Trends in annual and seasonal survival of pink-footed geese *Anser brachyrhynchus*. *Ibis* 144: 218–226.

Madsen, J. & Klaassen, M. 2006. Assessing body condition and energy budget components by scoring abdominal profiles in free-ranging pink-footed geese *Anser brachyrhynchus*. *Journal of Avian Biology* 37: 283–287.

Madsen, J., Tjørnløv, R. S., Frederiksen, M., Mitchell, C. & Sigfússon, A. T. 2014. Connectivity between flyway populations of waterbirds: assessment of rates of exchange, their causes and consequences. *Journal of Applied Ecology* 51: 183–193.

Madsen, J. & Williams, J. H. 2012. International species management plan for the Svalbard population of the pink-footed goose *Anser brachyrhynchus*. AEWA Technical Series No. 48. Bonn, Germany.

Manseau, M. & Gauthier, G. 1993. Interactions between greater snow geese and their rearing habitat. *Ecology* 74: 2045–2055.

Marshall, A. & Black, J. M. 1992. The effect of rearing experience on subsequent behaviour traits in captive-reared Hawaiian geese: implications for the re-introduction programme. *Bird Conservation International* 2: 131–147.

Martin, K., Cooch, E. G., Rockwell, R. F. & Cooke, F. 1985. Reproductive performance in lesser snow geese: are two parents essential? *Behavioral Ecology and Sociobiology* 17: 257–263.

Martin, P. & Bateson, P. 2007. *Measuring Behaviour: an introductory guide*. 3rd edn. Cambridge University Press, Cambridge.

Maynard Smith, J. 1982. *Evolution and the Theory of Games*. Cambridge University Press, Cambridge.

McKay, H. V., Bishop, J. D. & Ennis, D. C. 1994. The possible importance of nutritional requirements for dark-bellied brent geese in the seasonal shift from winter cereals to pasture. *Ardea* 82: 123–132.

McLandress, M. R. & Raveling, D. G. 1981. Hyperphagia and social behavior in Canada geese prior to spring migration. *Wilson Bulletin* 93: 310–324.

Mehlum, F. 1998. Areas in Svalbard important for geese during the pre-breeding, breeding and post-breeding periods. *Norsk Polarinstitutt Skrifter* 200: 41–56.

Meininger, P. L. 2002. Barnacle goose. In: *The Atlas of Breeding Birds in The Netherlands* (in Dutch). Naturalis, KNNV & European Invertebrate Survey – Nederland, Leiden, pp. 106–107.

Mennill, D. J., Doucet, S. M., Montomerie, R. & Ratcliffe, L. M. 2003. Achromatic color variation in black-capped chickadees, *Poecile atricapilla*: black and white signals of sex and rank. *Behavioral Ecology and Sociobiology* 53: 350–357.

Mini, A. & Black, J. M. 2009. Expensive traditions: energy expenditure of Aleutian geese in traditional and recently colonized habitats. *Journal of Wildlife Management* 73: 385–391.

Mini, A., Bachman, D., Cocke, J., Griggs, K., Spragens, K. A. & Black, J. M. 2011. Aleutian goose recovery: a 10-year review and future prospects. *Wildfowl* 61: 3–29.

Mitchell, C. R., Black, J. M. & Evans, M. 1998. Breeding success of cliff-nesting and island-nesting barnacle geese in Svalbard. *Norsk Polarinstitutt Skrifter* 200: 141–146.

Mock, D. W. & Parker, G. A. 1997. *The Evolution of Sibling Rivalry*. Oxford University Press, Oxford.

Moen, A. 1999. *National Atlas of Norway: Vegetation*. Norwegian Mapping Authority, Hønefoss.

Moody, A. T., Wilhelm, S. I., Cameron-MacMillan, M. L., Walsh, C. J. & Storey, A. E. 2005. Divorce in common murres (*Uria aalge*): relationship to parental quality. *Behavioral Ecology and Sociobiology* 57: 224–230.

Mortensen, C. E. 2011. Etablering og udvikling af ynglebestanden af bramgås på Saltholm, 1992–2010. *Dansk Ornitologisk Forenings Tidsskrift* 105: 159–166.

Moser, M. & Kalden, C. 1992. Farmers and waterfowl: conflict or coexistence – an introductory review. Recommendations. In: van Roomen, M. & Madsen, J. (eds) *Waterfowl and Agriculture: Review and future perspective of the crop damage conflict in Europe*. Proceedings of the international workshop 'Farmers and Waterfowl: Conflict or Coexistence', Lelystad, The Netherlands, 6–9 October 1991. IWRB Special Publication 21: 13–19.

Müller, M. J. 1980. *Handbuch Ausgewählter Klimastationen der Erde*. Gerold Richter, Universität Trier, Trier.

Newton, I. 1977. Timing and success of breeding in tundra nesting geese. In: Stonehouse, B. & Perrins, C. M. (eds) *Evolutionary Ecology*. Macmillan, London.

Newton, I. 1989. *Lifetime Reproduction in Birds*. Academic Press, London.

Newton, I. 1998. *Population Limitation in Birds*. Academic Press, San Diego.

Newton, I. 2004. Population limitation in migrants. *Ibis* 146: 197–226.

Nicolai, C. A., Sedinger, J. S., Ward, D. H. & Boyd, W. S. 2012. Mate loss affects survival but not breeding in black brant geese. *Behavioral Ecology* 23: 643–648.

Nilsson, J-Å. 1989. Causes and consequences of natal dispersal in the marsh tit, *Parus palustris. Journal of Animal Ecology* 58: 619–636.

Norderhaug, H. 1984. The Svalbard geese: an introductory review of research and conservation. *Norsk Polarinstitutt Skrifter* 181: 7–10.

Norris, K. & Evans, M. R. 2000. Ecological immunology: life history trade-offs and immune defense in birds. *Behavioral Ecology* 11: 19–26.

Ogilvie, M. A. 1972. Large numbered leg bands for individual identification of swans. *Journal of Wildlife Management* 36: 1261–1265.

Ogilvie, M. A. 1978. *Wild Geese.* T & AD Poyser, Calton.

Ogilvie, M. A., Boertmann, D., Cabot, D., Merne, O., Percival, S. M. & Sigfusson, A. 1999. Barnacle goose *Branta leucopsis*: Greenland. In: Madsen, J., Cracknell, G. & Fox A. D. (eds) *Goose Populations of the Western Palearctic.* Wetlands International, Wageningen, The Netherlands; and National Environmental Research Institute, Kalö, Denmark, pp. 246–256.

Ogilvie, M. A. & Owen, M. 1984. Some results from the ringing of barnacle geese *Branta leucopsis* in Svalbard and Britain. *Norsk Polarinstitutt Skrifter* 181: 49–55.

Otter, K. & Ratcliffe, L. 1996. Female initiated divorce in a monogamous songbird abandoning mates for males of higher quality. *Proceedings of the Royal Society B, Biological Sciences* 263: 351–354.

Oudman, T. 2009. Spring migration strategy and reproductive success in Svalbard barnacle geese *Branta leucopsis.* MSc Thesis, University of Groningen.

Ouweneel, G. L. 2001. Rapid growth of a breeding population of barnacle geese *Branta leucopsis* in the Delta area, SW-Netherlands (in Dutch). *Limosa* 74: 137–146.

Owen, M. 1971. The selection of feeding site by white-fronted geese in winter. *Journal of Applied Ecology* 8: 905–917.

Owen, M. 1972a. Some factors affecting food intake and selection in white-fronted geese. *Journal of Animal Ecology* 41: 71–92.

Owen, M. 1972b. Movements and feeding ecology of white-fronted geese at the New Grounds, Slimbridge. *Journal of Applied Ecology* 9: 385–398.

Owen, M. 1973. The winter feeding ecology of wigeon at Bridgwater Bay, Somerset. *Ibis* 115: 227–243.

Owen, M. 1975. An assessment of the fecal analysis technique in waterfowl feeding studies. *Journal of Wildlife Management* 39: 271–279.

Owen, M. 1976a. Factors affecting the distribution of geese in the British Isles. *Wildfowl* 27: 143–147.

Owen, M. 1976b. The election of winter food by white-fronted geese. *Journal of Applied Ecology* 13: 715–72.

Owen, M. 1977. The role of wildfowl refuges on agricultural land in lessening the conflict between farmers and geese in Britain. *Biological Conservation* 11: 209–222.

Owen, M. 1978. Food selection in geese. *Verhandlungen Ornithologischen Gesellschaft in Bayern* 23: 169–176.

Owen, M. 1980a. *Wild Geese of the World.* Batsford, London.

Owen, M. 1980b. The role of refuges in wildfowl management. In: Wright, E. N., Feare, C. J. & Inglis, I. R. (eds) *Bird Problems in Agriculture.* British Crop Protection Council, London, pp. 44–61.

Owen, M. 1981a. Food selection in geese. *Verhandlungen der Ornithologischen Gesellschaft in Bayern* 23: 169–176.

Owen, M. 1981b. Abdominal profile – a condition index for wild geese in the field. *Journal of Wildlife Management* 45: 227–230.

Owen, M. 1982. Population dynamics of Svalbard barnacle geese, 1970–1980. The rate, pattern and causes of mortality as determined by individual marking. *Aquila* 89: 229–247.

Owen, M. 1984. Dynamics and age structure of an increasing goose population – the Svalbard barnacle goose *Branta leucopsis*. *Norsk Polarinstitutt Skrifter* 181: 37–47.

Owen, M. 1990a. *The Barnacle Goose*. Shire Natural History, 51. Aylesbury, Buckinghamshire.

Owen, M. 1990b. The damage-conservation interface illustrated by geese. *Ibis* 132: 238–252.

Owen, M. & Black, J. M. 1989a. Factors affecting the survival of barnacle geese on migration from the wintering grounds. *Journal Animal Ecology* 58: 603–618.

Owen, M. & Black, J. M. 1989b. Barnacle goose. In: Newton, I. (ed) *Lifetime Reproduction in Birds*. Academic Press, London, pp. 349–362.

Owen, M. & Black, J. M. 1990. *Waterfowl Ecology*. Blackie Publishers, Glasgow.

Owen, M. & Black, J. M. 1991a. The importance of migration mortality in non-passerine birds. In: Perrins, C. M., Lebreton, J-D. & Hirons, G. J. M. (eds) *Bird Population Studies: relevance to conservation and management*. Oxford University Press, Oxford, pp. 360–372.

Owen, M. & Black, J. M. 1991b. Geese and their future fortunes. *Ibis* 133: S28–S35.

Owen, M. & Black, J. M. 1999. Barnacle goose *Branta leucopsis*: Svalbard. In: Madsen, J., Cracknell, G. & Fox, A. D. (eds) *Goose Populations of the Western Palearctic*. Wetlands International, Wageningen, The Netherlands; and National Environmental Research Institute, Kalö, Denmark, pp. 258–269.

Owen, M., Black, J. M., Agger, M. C. & Campbell, C. R. G. 1987. The use of the Solway Firth by an increasing population of barnacle geese in relation to changes in refuge management. *Biological Conservation* 39: 63–81.

Owen, M., Black, J. M. & Liber, H. 1988. Pair bond duration and the timing of its formation in barnacle geese. In: Weller, M. (ed) *Waterfowl in Winter*. University of Minnesota Press, Minneapolis, pp. 23–38.

Owen, M. & Gullestad, N. 1984. Migration routes of Svalbard barnacle geese *Branta leucopsis* with a preliminary report on the importance of Bjørnøya staging area. *Norsk Polarinstitutt Skrifter* 181: 67–77.

Owen, M. & Kerbes, R. H. 1971. On the autumn food of barnacle geese at Caerlaverock National Nature Reserve. *Wildfowl* 22: 114–119.

Owen, M. & Norderhaug, M. 1977. Population dynamics of barnacle geese *Branta leucopsis* breeding in Svalbard, 1948–1976. *Ornis Scandinavica* 8: 161–174.

Owen, M. & Ogilvie, M. A. 1979. Wing molt and weights of barnacle geese in Spitsbergen. *Condor* 81: 42–52.

Owen, M. & Shimmings, P. 1992. The occurrence and performance of leucistic barnacle geese *Branta leucopsis*. *Ibis* 134: 22–26.

Owen, M. & Wells, R. 1979. Territorial behaviour of breeding geese – a re-examination of Ryder's hypothesis. *Wildfowl* 30: 20–26.

Owen, M., Wells, R. L. & Black, J. M. 1992. Energy budgets of wintering barnacle geese: The effects of declining food resources. *Ornis Scandinavica* 23: 451–458.

Partridge, L. 1989. Lifetime reproductive success and life-history evolution. In: Newton, I. (ed) *Lifetime Reproduction in Birds*. Academic Press, London, pp. 421–440.

Patterson, D. J. 1995. Raptor evasion and defensive behaviour by barnacle geese. *Scottish Birds* 18: 101–102.

Patterson, D. J. 1998. Mobbing behaviour by barnacle geese on a ground predator. *Scottish Birds* 19: 168–169.

Pauliny, A., Larsson, K. & Blomqvist, D. 2012. Telomere dynamics in a long-lived bird, the barnacle goose. *BMC Evolutionary Biology* 12: 257.

Paxinos, E. E., James, H. F., Olson, S. L., Sorenson, M. D., Jackson, J. & Fleisher, R.C. 2002. mtDNA from fossils reveals a radiation of Hawaiian geese recently derived from the Canada goose *Branta canadensis*. *Proceedings of the National Academy of Sciences* 99: 1399–1404.

Pennycuick, C. J. 1989. *Bird Flight Performance: a practical calculation manual.* Oxford University Press, Oxford.

Percival, S. M. 1991. The population structure of Greenland barnacle geese *Branta leucopsis* on the wintering grounds on Islay. *Ibis* 133: 357–364.

Perrins, C. M. 1970. The timing of birds' breeding seasons. *Ibis* 112: 242–255.

Person, B. T., Herzog, M. P., Ruess, R. W., Sedinger, J. S., Anthony, R. M. & Babcock, C. A. 2003. Feedback dynamics of grazing lawns: coupling vegetation change with animal growth. *Oecologia* 135: 583–592.

Petrie, M. & Møller, A. P. 1991. Laying eggs in others' nests: Intraspecific brood parasitism in birds. *Trends in Ecology and Evolution* 6: 315–320.

Pettifor, R. A., Black, J. M., Owen, M., Rowcliffe, J. M. & Patterson, D. 1998. Growth of the Svalbard barnacle goose *Branta leucopsis* winter population 1958–1996: An initial review of temporal demographic changes. *Norsk Polarinstitutt Skrifter* 200: 147–164.

Pettifor, R. A., Caldow, R. W. G., Rowcliffe, J. M., Goss-Custard, J. D., Black, J. M., Hodder, K. H., Houston, A. I., Lang, A. & Webb, J. 2000. Spatially explicit, individual-based, behavioural models of the annual cycle of two migratory goose populations – model development, theoretical insights and applications. *Journal of Applied Ecology* 37: 103–135.

Phillips, R. A., Cope, D. R., Rees, E. C., O'Connell, M. J. 2003. Site fidelity and range size of wintering barnacle geese *Branta leucopsis. Bird Study* 50: 161–169.

Piersma, T. & Baker, A. J. 2000. Life history characteristics and the conservation of migratory shorebirds. In: Gosling, L. M. & Sutherland, W. J. (eds) *Behaviour and Conservation.* Cambridge University Press, Cambridge, pp. 105–124.

Porter, A. H. & Johnson, N. A. 2002. Speciation despite gene flow when developmental pathways evolve. *Evolution* 56: 2103–2111.

Prestrud, P., Black, J. M. & Owen, M. 1989. The relationship between an increasing population of barnacle geese and the number and size of their colonies in Svalbard. *Wildfowl* 40: 32–38.

Prestrud, P. & Børset, A. 1984. Status of the goose populations in the bird sanctuaries in Svalbard. *Norsk Polarinstitutt Skrifter* 181: 129–134.

Prins, H. H. T. & Ydenberg, R. C. 1985. Vegetation growth and a seasonal habitat shift of the barnacle goose (*Branta leucopsis*). *Oecologia* 66: 122–125.

Prop, J. 1991. Food exploitation patterns by brent geese *Branta bernicla* during spring staging. *Ardea* 79: 331–342.

Prop, J. 2004. Food finding: on the trail to successful reproduction in migratory geese. PhD Thesis. University of Groningen, Groningen.

Prop, J. & Black, J. M. 1998. Food intake, body reserves and reproductive success of barnacle geese *Branta leucopsis* staging in different habitats. *Norsk Polarinstitutt Skrifter* 200: 175–193.

Prop, J., Black, J. M. & Shimmings, P. 2003. Travel schedules to the high arctic: barnacle geese trade-off the timing of migration with accumulation of fat deposits. *Oikos* 103: 403–414.

Prop, J., Black, J. M., Shimmings, P. & Owen, M. 1998. The spring range of barnacle geese *Branta leucopsis* in relation to changes in land management and climate. *Biological Conservation* 86: 339–346.

Prop, J. & de Fouw, J. 2004. *Reproductive Output of Barnacle Geese on Nordenskiöldkysten, Svalbard, Summer 2004.* Report, University Groningen, Groningen.

Prop, J. & de Vries, J. 1993. Impact of snow and food conditions on the reproductive performance of barnacle geese *Branta leucopsis. Ornis Scandinavica* 24: 110–121.

Prop, J. & Deerenberg, C. 1991. Spring staging in brent geese *Branta bernicla*: feeding constraints and the impact of diet on the accumulation of body reserves. *Oecologia* 87: 19–28.

Prop, J. & Drent, R. H. 2003. *Goose Census of Nordenskiöldkysten, West-Spitsbergen, Svalbard, Summer 2003.* Report, University Groningen, Groningen.

Prop, J., Drent, R. H. & Owen, M. 2004. Survival costs related to the timing of breeding and brood size in Arctic barnacle geese. In: Prop, J. (ed) Food finding: on the trail to successful reproduction in migratory geese. PhD Thesis, University of Groningen, Groningen, pp. 213–229.

Prop, J. & Loonen, M. 1989. Goose flocks and food exploitation: the importance of being first. *Acta XIX Congressus Internationalis Ornithologici*: 1878–1887.

Prop, J., Oudman, T., van Spanje, T. M. & Wolters, E. H. 2013. Patterns of predation of pink-footed goose nests by polar bear. *Ornis Norvegica* 36: 38–46.

Prop, J. & Quinn, J. L. 2004. Interference competition, foraging routines and reproductive success in red-breasted geese: fight or flight? In: Prop, J. (ed) Food finding: on the trail to successful reproduction in migratory geese. PhD Thesis, University of Groningen, Groningen, pp. 153–169.

Prop, J., van Eerden, M. R., Daan, S., Drent, R. H., Tinbergen, J. M. & St Joseph, A. M. 1980. Ecology of the barnacle goose (*Branta leucopsis*) during the breeding season – Preliminary results from expeditions to Spitsbergen in 1977 and 1978. In: *Proceedings of the Norwegian-Netherlands Symposium on Svalbard*. Arctic Centre, Groningen, pp. 50–112.

Prop, J., van Eerden, M. R. & Drent, R. H. 1984. Reproductive success of the barnacle goose in relation to food exploitation on the breeding grounds, western Spitsbergen. *Norsk Polarinstitutt Skrifter* 181: 87–117.

Prop, J., van Marken Lichtenbelt, W. D., Beekman, J. H. & Faber, J. F. 2005. Using food quality and retention time to predict digestion efficiency in geese. *Wildlife Biology* 11: 21–29.

Prop, J. & Vulink, T. 1992. Digestion by barnacle geese in the annual cycle: the interplay between retention time and food quality. *Functional Ecology* 6: 180–189.

Pulliam, H. R. & Caraco, T. 1984. Living in groups: is there an optimal group size? In: Krebs, J. R. & Davies, N. B. (eds) *Behavioural Ecology: An evolutionary approach*. 2nd edn. Blackwell Scientific Publications, Oxford, pp. 122–147.

Quinn, J. L., Prop, J., Kokorev, Y. & Black, J. M. 2003. Predator protection or similar habitat in red-breasted goose nesting associations: extremes along a continuum. *Animal Behaviour* 65: 297–307.

Radesäter, T. 1974. Form and sequential associations between the triumph ceremony and other behaviour patterns in the Canada goose *Branta canadensis* L. *Ornis Scandinavica* 5: 87–101.

Ranta, E., Lindström, J. & Lindén, H. 1995. Synchrony in tetraonid population dynamics. *Journal of Animal Ecology* 64: 767–776.

Raveling, D. G. 1969a. Preflight and flight behavior in Canada geese. *Auk* 86: 671–681.

Raveling, D. G. 1969b. Roost sites and flight patterns in Canada geese in winter. *Journal of Wildlife Management* 33: 319–330.

Raveling, D. G. 1970. Dominance relationships and agonistic behaviour of Canada geese in winter. *Behaviour* 37: 291–319.

Raveling, D. G. 1978. Dynamics and distribution of Canada geese in winter. *Proceedings of the North American Wildlife Society Conference* 43: 206–225.

Raveling, D. G. 1979. Traditional use of migration and winter roost sites by Canada geese. *Journal of Wildlife Management* 43: 229–235.

Raveling, D. G. 1981. Survival, experience, and age in relation to breeding success of Canada geese. *Journal of Wildlife Management* 45: 817–829.

Raveling, D. G., Sedinger, J. S. & Johnson, D. S. 2000. Reproductive success and survival in relation to experience during the first two years in Canada geese. *Condor* 102: 941–945.

Ree, V. 2001. Direktoratet for naturforvaltning med mangelfull informasjon om hvitkinngåsa i Sørøst-Norge. http://www.birdlife.no/naturforvaltning/nyheter/?id=373 Accessed January 2014

Rees, E. C. 2006. *Bewick's Swan*. T & AD Poyser, London.

Rees, E. C. 1987. Conflict of choice among pairs of Bewick's swans regarding their migratory movements to and from the wintering grounds. *Animal Behaviour* 35: 1685–1693.

Rees, E. C., Lievesley, P., Pettifor, R. A. & Perrins, C. 1996. Mate fidelity in swans: an interspecific comparison. In: Black, J. M. (ed) *Partnerships in Birds. The Study of Monogamy.* Oxford University Press, Oxford, pp. 118–137.

Rees, E. C., Owen, M., Gitay, H. & Warren, S. 1990. The fate of plastic leg rings on geese and swans. *Wildfowl* 41: 43–52.

Remmert, H. 1980. *Arctic Animal Ecology.* Springer Verlag, Berlin.

Ricklefs, R. E. 2000. Intrinsic aging-related mortality in birds. *Journal of Avian Biology* 31: 103–111.

Ridgill, S. C. & Fox, A. D. 1990. Cold weather movements of waterfowl in Western Europe. *IWRB Special Publication* 13, Slimbridge, UK.

Robertson, D. G. & Slack, D. R. 1995. Landscaping change and its effects on the wintering range of a lesser snow goose *Chen caerulescens caerulescens* population: a review. *Biological Conservation* 71: 179–185.

Rockwell, R. F. & Gormezano, L. J. 2009. The early bear gets the goose: climate change, polar bears and lesser snow geese in western Hudson Bay. *Polar Biology* 32: 539–547.

Rodway, M. S. 2007. Timing of pairing in waterfowl I: reviewing the data and extending the theory. *Waterbirds* 30: 488–505.

Rohwer, F. C. & Anderson, M. G. 1988. Female-biased philopatry, monogamy, and the timing of pair formation in migratory waterfowl. In: Johnston, R. F. (ed) *Current Ornithology.* Plenum Press, New York, pp. 187–221.

Rohwer, F. C. & Freeman, S. 1989. The distribution of conspecific nest parasitism in birds. *Canadian Journal of Zoology* 67: 239–253.

Rohwer, S. 1985. Dyed birds achieve higher social status than controls in Harris' sparrows. *Animal Behaviour* 33: 1325–1331.

Rowcliffe, M., Pettifor, R. A. & Carbone, C. 2004. Foraging inequalities in large groups: quantifying depletion experienced by individuals in goose flocks. *Journal of Animal Ecology* 73: 97–108.

Rowcliffe, J. M., Watkinson, A. R., Sutherland, W. J. & Vickery, J. A. 1995. Cyclic winter grazing patterns in brent geese and the regrowth of salt-marsh grass. *Functional Ecology* 9: 931–941.

Rowcliffe, J. M., Watkinson, A. R., Sutherland, W. J. & Vickery, J. A. 2001. The depletion of algal beds by geese: A predictive model and test. *Oecologia* 127: 361–371.

Ruokonen, M. 2001. Phylogeography and conservation genetics of the lesser white-fronted goose (*Anser erythropus*). Dissertation, University of Oulu. http://herkules.oulu.fi/isbn9514259483 (November 2006).

SAS Institute (1990–2001) *Version 6–8* edn. SAS Institute, Cary, North Carolina.

Saino, N., Ambrosini, R., Rubolini, D., von Hardenberg, J., Provenzale, A., Hüppop, K., Hüppop, O., Lehikoinen, A., Lehikoinen, E., Rainio, K., Romano, M. & Sokolov, L. 2011. Climate warming, ecological mismatch at arrival and population decline in migratory birds. *Proceedings of the Royal Society B, Biological Sciences* 278: 835–842.

Sandberg, R. & Moore, F. R. 1996. Fat stores and arrival on the breeding grounds: reproductive consequences for passerine migrants. *Oikos* 77: 577–581.

Scheiber, I. B. R., Kotrschal, K. & Weiß, B. M. 2009. Benefits of family reunions: Social support in secondary greylag goose families. *Hormones and Behavior* 55: 133–138.

Scheiber, I. B. R., Weiß, B. M., Hemetsberger, J. & Kotrschal, K. 2013. *The Social Life of Greylag Geese: Patterns, mechanisms and evolutionary function in an avian model system.* Cambridge University Press, Cambridge.

Schindler, M. & Lamprecht, J. 1987. Increase in parental effort with brood size in a nidifugous bird. *Auk* 104: 688–693.

Schmutz, J. A. & Ely, C. R. 1999. Survival of greater white-fronted geese: effects of year, season, sex, and body condition. *Journal of Wildlife Management* 63: 1239–1248.

Schwabl, H. 1993. Yolk is a source of maternal testosterone for developing birds. *Proceedings of the National Academy of Sciences* 90: 11446–11450.

Scott, D. A. 1980. A preliminary inventory of wetlands of international importance for waterfowl in west Europe and northwest Africa. *IWRB Special Publication* 2, IWRB, Slimbridge.

Scott, D. K. 1980a. Functional aspects of prolonged parental care in Bewick's swans. *Animal Behaviour* 28: 938–952.

Scott, D. K. 1980b. Functional aspects of the pair bond in Bewick's swans. *Behavioral Ecology and Sociobiology* 7: 323–327.

Scott, P. 1938. *Wild Chorus*. Country Life, London.

Scott, P. 1961. *The Eye of the Wind*. Hodder and Stoughton, London.

Scott, P. 1981. *Observations of Wildlife*. Phaidon Press, Oxford.

Scott, P. & Fisher, J. 1953. *A Thousand Geese*. Collins, London.

Sedinger, J. S., Chelgren, N. D., Lindberg, M. S., Obritchkewitch, T., Kirk, M. T., Martin, P., Anderson, B. A. & Ward, D. H. 2002. Life-history implications of large-scale spatial variation in adult survival of black brant (*Branta bernicla nigricans*). *Auk* 119: 510–515.

Sedinger, J. S., Chelgren, N. D., Ward, D. H. & Lindberg, M. S. 2008. Fidelity and breeding probability related to population density and individual quality in black brent geese *Branta bernicla nigricans*. *Journal of Animal Ecology* 77: 702–712.

Sedinger, J. S. & Flint, P. L. 1991. Growth rate is negatively correlated with hatch date in black brant. *Ecology* 72: 496–502.

Sedinger, J. S., Flint, P. L. & Lindberg, M. S. 1995. Environmental influence on life-history traits: growth, survival and fecundity in black brant (*Branta bernicla*). *Ecology* 76: 2404–2414.

Sedinger, J. S. & Raveling, D. G. 1984. Dietary selectivity in relation to availability and quality of food for goslings of cackling geese. *Auk* 101: 295–306.

Sedinger, J. S. & Raveling, D. G. 1986. Timing of nesting by Canada geese in relation to the phenology and availability of their food plants. *Journal of Animal Ecology* 55: 1083–1102.

Sedinger, J. S. & Raveling, D. G. 1988. Foraging behavior of cackling Canada goose goslings: implications for the roles of food availability and processing rate. *Oecologia* 75: 119–124.

Sedinger, J. S. & Raveling, D. G. 1990. Parental behavior of cackling Canada geese during brood rearing: division of labor within pairs. *Condor* 92: 174–181.

Sedinger, J. S., Schamber, J. L., Ward, D. H., Nicolai, C. A. & Conant, B. 2011. Carryover effects associated with winter location affect fitness, social status, and population dynamics in a long-distance migrant. *American Naturalist* 178: E110–E123.

Sedinger, J. S., Ward, D. H., Schamber, J. L., Butler, W. I., Eldridge, W. D., Conant, B., Voelzer, J. E., Chelgren, N. D. & Herzog, M. P. 2006. Effects of El Niño on distribution and reproductive performance of black brant. *Ecology* 87: 151–159.

Shaughnessy, F. J., Gilkerson, G., Black, J. M., Ward, D. H. & Petrie, M. 2012. Predicted *Zostera marina* response to sea level rise and availability to foraging black brant in Pacific coast estuaries. *Ecological Applications* 22: 1743–1761.

Sherman, P. W. 2001. Wood ducks: a model system for investigating conspecific parasitism in cavity-nesting birds. In: Dugatkin, L. A. (ed) *Model Systems in Behavioural Ecology*. Princeton University Press, Princeton, pp. 309–335.

Shields, W. M. 1984. Optimal inbreeding and the evolution of philopatry. In: Swingland, I. R. & Greenwood, P. J. (eds) *The Ecology of Animal Movement*. Clarendon Press, Oxford, pp. 132–159.

Shimmings, P. 2003. *Spring Staging by Barnacle Geese* Branta leucopsis, *and the Effects of a Management Plan in the Herøy District in Nordland, Norway*. Report to Direktoratet for naturforvaltning, Herøy kommune, Nordland.

Shimmings, P. & Isaksen, K. 2006. *Results of Fieldwork on Barnacle Geese* Branta leucopsis *during Spring Migration along the Norwegian Coast in 2006.* Report to Fylkesmannen i Nordland, Miljøvernavdelingen, Direktoratet for naturforvaltning, Herøy kommune, Nordland.

Shimmings, P. & Isaksen, K. 2013. *Overvåking av rastende hvitkinngjess Branta leucopsis langs norskekysten våren 2013.* Rapport til Fylkesmannen i Nordland og Herøy kommune,,Nordland.

Shipley, L. A., Illius, A. W., Danell, K., Hobbs, N.T. & Spalinger, D. E. 1999. Predicting bite size selection of mammalian herbivores: A test of a general model of diet optimization. *Oikos* 84: 55–68.

Sibly, R. M. & Hone, J. 2003. Population growth rate and its determinants: an overview. In: Sibly, R. M., Hone, J. & Clutton-Brock, T. H. (eds) *Wildlife Population Growth Rates.* Cambridge University Press, Cambridge.

Siriwardena, G. M. & Black, J. M. 1999. Parent and gosling strategies in wintering barnacle geese. *Wildfowl* 49: 18–26.

Slatkin, M. 1987. Gene flow and the geographic structure of natural populations. *Science* 236: 787–792.

Spaans, B., Blijleven, H. J., Popov, I. U., Rykhlikova, M. E. & Ebbinge, B. S. 1998. Dark-bellied brent geese *Branta bernicla bernicla* forego breeding when Arctic foxes *Alopex lagopus* are present during nest initiation. *Ardea* 86: 11–20.

Spoon, T. R., Millam, J. R. & Owings, D. H. 2004. Variation in the stability of cockatiel (*Nymphicus hollandicus*) pair relationships: the roles of males, females, and mate compatibility. *Behaviour* 141: 1211–1234.

Stamps, J. A. & Groothius, T. G. G. 2010. Developmental perspectives on personality: implications for ecological and evolutionary studies of individual differences. *Philosophical Transactions of the Royal Society B* 365: 4029–4041.

Stearns, S. C. 1992. *The Evolution of Life Histories.* Oxford University Press, Oxford.

Stempniewicz, L. 2006. Polar bear predatory behaviour towards molting barnacle geese and nesting glaucous gulls on Spitsbergen. *Arctic* 59: 247–251.

Stillman, R. A. & Goss-Custard, J. D. 2010. Individual-based ecology of coastal birds. *Biological Reviews* 85: 413–434.

Stirling, I. 2011. *Polar Bears: the natural history of a threatened species.* Fitzhenry & Whiteside, Markham, Canada.

Sutherland, W. J. 1996. *From Individual Behaviour to Population Ecology.* Oxford University Press, Oxford.

Sutherland, W. J. & Allport, G. A. 1994. A spatial depletion model of the interaction between bean geese and wigeon with the consequences for habitat management. *Journal of Animal Ecology* 63: 51–59.

Swingland, I. R. & Greenwood, P. J. 1984. *The Ecology of Animal Movement.* Clarendon Press, Oxford.

Teunissen, W., Spaans, B. & Drent, R. H. 1985. Breeding success in the brent in relation to individual feeding opportunities during spring staging in the Wadden Sea. *Ardea* 73: 109–119.

Therkildsen, O. R. & Madsen, J. 2000. Assessment of food intake rates in pink-footed geese *Anser brachyrhynchus* based on examination of oesophagus contents. *Wildlife Biology* 6: 167–172.

Tombre, I. M. & Erikstad, K. E. 1996. An experimental study of incubation effort in high-Arctic barnacle geese. *Journal of Animal Ecology* 65: 325–331.

Tombre, I. M., Erikstad, K. E. & Bunes, V. 2012. State-dependent incubation behaviour in the high arctic barnacle geese. *Polar Biology* 35: 985–992.

Tombre, I. M., Eythórsson, E. & Madsen, J. 2013a. Towards a solution to the goose-agriculture conflict in north Norway, 1988–2012: The interplay between policy, stakeholder influence and goose population dynamics. *PLoS ONE* 8: e71912.

Tombre, I. M., Eythórsson, E. & Madsen, J. 2013b. Stakeholder involvement in adaptive goose management: case studies and experiences from Norway. *Ornis Norvegica* 36: 17–24.

Tombre, I. M., Hogda, K. A., Madsen, J., Griffin, L. R., Kuijken, E., Shimmings, P., Rees, E. & Verscheure, C. 2008. The onset of spring and timing of migration in two arctic nesting goose populations: the pink-footed goose *Anser brachyrhynchus* and the barnacle goose *Branta leucopsis*. *Journal of Avian Biology* 39: 691–703.

Tombre, I. M., Madsen, J., Tommervik, H., Haugen, K. P. & Eythorsson, E. 2005. Influence of organised scaring on distribution and habitat choice of geese on pastures in Northern Norway. *Agriculture Ecosystems & Environment* 111: 311–320.

Tombre, I. M., Mehlum, F. & Loonen, M. J. J. E. 1998. The Kongsfjorden colony of barnacle geese: nest distribution and the use of breeding islands, 1980–1997. *Norsk Polarinstitutt Skrifter* 200: 57–66.

Torres Esquivias, J. A. & Ayala Moreno, J. M. 1986. Variation du dessein cephalique des males de l'erismature à tête blanche (*Oxyura leucocephala*). *Alauda* 54: 197–206.

Trivers, R. L. 1972. Parental investment and sexual selection. In: Campbell, B. (ed) *Sexual Selection and the Descent of Man*. Aldine, Chicago, pp. 136–179.

Trost, R., Dickson, K. & Zavaleta, D. 1993. Harvesting waterfowl on a sustained yield basis: the North American perspective. *IWRB Special Publication* 26: 106–112.

Turchin, P. 2003. *Complex Population Dynamics: a theoretical/empirical synthesis*. Princeton University Press, Princeton.

Turcotte, Y. & Bédard, J. 1989. Prolonged parental care and foraging of greater snow goose juveniles. *Wilson Bulletin* 101: 500–503.

Tyrberg, T. 1998. Pleistocene birds of the Palearctic: A catalogue. Publications of the Nuttall Ornithological Club no 27, Cambridge, MA, 270 pp; with supplements. http://w1.115.telia.com/~u11502098/pleistocene.html Accessed January 2014.

United States Department of Agriculture (USDA) 2014. Farm bill resources. http://www.ers.usda.gov/farm-bill-resources.aspx#.UxFpFfl92So Accessed February 2014.

United States Department of the Interior and Environment Canada 1986. *North American Waterfowl Management Plan*. Washington, D.C.

Väänänen, V.-M., Laine, J., Lammi, E., Lehtiniemi, T., Luostarinen, V-M. & Mikkola-Roos, M. 2010. The establishment of barnacle goose in Finland – rapid growth rate and expansion of the breeding grounds. *Linnut-vuosikirja* 2009: 72–77.

Väänänen, V.-M., Nummi, P., Lehtiniemi, T., Luostarinen, V-M. & Mikkola-Roos, M. 2011. Habitat complementation in urban barnacle geese: from safe nesting islands to productive foraging lawns. *Boreal Environment Research* 16 (suppl. B): 26–34.

van de Pol, M., Bruinzeel, L. W., Heg, D., van der Jeurgd, H. P. & Verhulst, S. 2006a. A silver spoon for a golden future: long-term effects of natal origin on fitness prospects of oystercatchers (*Haematopus ostralegus*). *Journal of Animal Ecology* 75: 616–626.

van de Pol, M., Heg, D., Bruinzeel, L. W., Kuijper, B. & Verhulst, S. 2006b. Experimental evidence for a causal effect of pair-bond duration on reproductive performance in oystercatchers (*Haematopus ostralegus*). *Behavioral Ecology* 17: 982–991.

van der Graaf, A. J., Stahl, J., Klimkowska, A., Bakker, J. P. & Drent, R. H. 2006. Surfing on a green wave – how plant growth drives spring migration in the barnacle goose *Branta leucopsis*. *Ardea* 94: 567–577.

van der Jeugd, H. P. 2001. Large barnacle goose *Branta leucopsis* males can overcome the social costs of natal dispersal. *Behavioral Ecology* 12: 275–282.

van der Jeugd, H. P. 2013. Survival and dispersal in a newly-founded temperate barnacle goose *Branta leucopsis* population. *Wildfowl* 63: 72–89.

van der Jeugd, H. P. & Blaakmeer, K. B. 2001. Teenage love: the importance of trial liaisons, subadult plumage and early pairing in barnacle geese. *Animal Behaviour* 62: 1075–1083.

van der Jeugd, H. P., Eichhorn, G., Litvin, K. E., Stahl, J., Larsson, K., van der Graaf, A. J. & Drent, R. H. 2009. Keeping up with early springs: rapid range expansion in an avian herbivore incurs a mismatch between reproductive timing and food supply. *Global Change Biology* 15: 1057–1071.

van der Jeugd, H. P. & Larsson, K. 1998. Pre-breeding survival of barnacle geese *Branta leucopsis* in relation to fledgling characteristics. *Journal of Animal Ecology* 67: 953–966.

van der Jeugd, H. P. & Larsson, K. 1999. Density-dependent effects on age at first reproduction and natal dispersal in the barnacle goose *Branta leucopsis*. In: van der Jeugd, H. P. (ed) Life history decisions in a changing environment. PhD Thesis, Uppsala University, Uppsala, Sweden, pp. 72–92.

van der Jeugd, H. P. & Litvin, K. Y. 2006. Travels and traditions: long-distance dispersal in the Barnacle Goose *Branta leucopsis* based on individual case histories. *Ardea* 94: 421–432.

van der Jeugd, H. P., Olthoff, M. P. & Stahl, J. 2001. Breeding range translates into staging site choice: Baltic and Arctic barnacle geese *Branta leucopsis* use different habitats at a Dutch Wadden Sea Island. *Ardea* 89: 253–265.

van der Jeugd, H. P., van der Veen, I. T. & Larsson, K. 2002. Kin clustering in barnacle geese: familiarity or phenotype matching? *Behavioral Ecology* 13: 786–790.

van der Jeugd, H. P., van Winden, E. & Koffijberg, K. 2008. *Evaluatie Opvangbeleid 2005–2008 overwinterende ganzen en smienten, deelrapport 5: Invloed opvangbeleid op de verspreiding van overwinterende ganzen en smienten binnen Nederland.* SOVON-report 2008/20. SOVON Vogelonderzoek Nederland, Beek-Ubbergen.

van der Leeuw, A. H. J., Kurk, K., Snelderwaard, P. C., Bout, R. G. & Berkhoudt, H. 2003. Conflicting demands on the trophic system of Anseriformes and their evolutionary implications. *Animal Biology* 53: 259–301.

van der Wal, R., Madan, N., van Lieshout, S., Dormann, C., Langvatn, R. & Albon, S. D. 2000. Trading forage quality for quantity? Plant phenology and patch choice by Svalbard reindeer. *Oecologia* 123: 108–115.

van der Wal, R., van de Koppel, J. & Sagel, M. 1998. On the relation between herbivore foraging efficiency and plant standing crop: an experiment with barnacle geese. *Oikos* 82: 123–130.

van Eerden, M. R. 1997. *Patchwork – Patch use, habitat exploitation and carrying capacity for water birds in Dutch freshwater wetlands.* Van Zee tot Land 65, Lelystad, The Netherlands.

van Roomen, M. J. W., van Winden, E. A. J., Koffijberg, K., Kleefstra, R., Ottens, G., Voslamber, B. & SOVON Ganzen- en zwanenwerkgroep. 2003. *Waterbirds in The Netherlands in 2001/2002.* SOVON, Beek-Ubbergen.

Vickery, J. A., Watkinson, A. R. & Sutherland, W. J. 1994. The solutions to the brent goose problem: an economic analysis. *Journal of Applied Ecology* 31: 371–382.

Vine, D. A. 1983. Sward structure changes within a perennial ryegrass sward: leaf appearance and death. *Grass and Forage Science* 38: 231–242.

Ward, D. H., Rexstad, E. A., Sedinger, J. S., Lindberg, M. S. & Dawe, N. K. 1997. Seasonal and annual survival of adult Pacific brant. *Journal of Wildlife Management* 61: 773–781.

Warren, S. M., Fox, A. D. & Walsh, A. 1993. Extended parent-offspring relationships amongst the Greenland white-fronted goose *Anser albifrons flavirostris*. *Auk* 110: 145–148.

Welsh, D. & Sedinger, J. S. 1990. Extra-pair copulations in black brant. *Condor* 92: 242–244.

White, C. M., Clum, N. J., Cade, T. J. & Hunt, W. G. 2002. Peregrine falcon (*Falco peregrinus*). In: Poole, A. & Gill, F. (eds) *Birds of North America*. The Academy of Natural Sciences, Philadelphia, and The American Ornithologists' Union, Washington, D.C. 660: 1–48.

White, G. C. & Burnham, K. P. 1999. Program MARK, survival estimation from populations of marked animals. *Bird Study* 46: 120–138.

Whitford, P. C. 1998. Vocal and visual communication of giant Canada geese. In: Rusch, D. H., Samuel, M. D., Humburg,

D. D. & Sullivan, B. D. (eds) *Biology and Management of Canada Geese.* Proceedings of the International Canada Goose Symposium, Milwaukee, Wisconsin, pp. 375–386.

Wibeck, E. 1946. De vitkindade gässen vid Gotlandskusten. In: Pettersson, B. & Curry-Lindahl, K. (eds). *Natur på Gotland Svensk Natur.* Stockholm, pp. 253–258.

Wildfowl and Wetlands Trust (WWT) 2013. WWT waterbird monitoring. Barnacle goose *Branta leucopsis.* http://monitoring. wwt.org.uk/our-work/goose-swan-monitoring-programme/species-accounts/barnacle-goose/ Accessed January 2014.

Williams, B. K., Koneff, M. D. & Smith, D. A. 1999. Evalutation of waterfowl conservation under the North American Waterfowl Management plan. *Journal of Wildlife Management* 63: 417–440.

Williams, T. D., Cooch, E. G., Jefferies, R. L. & Cooke, F. 1993. Environmental degradation, food limitation and reproductive output: juvenile survival

in lesser snow geese. *Journal of Animal Ecology* 62: 766–777.

Wittenberger, J. F. 1983. Tactics of mate choice. In: Bateson, P. (ed) *Mate Choice.* Cambridge University Press, Cambridge, pp. 435–447.

Wold, H. A. 1985. *Utvær* (Outer-island): *Pictures from an everyday landscape in northern Norway* (in Norwegian). J. W. Cappelens Forlag, Oslo.

Ydenberg, R. C., Prins, H. H. T. & van Dijk, J. 1983. The post-roost gatherings of wintering Barnacle Geese: Information centres? *Ardea* 71: 125–131.

Yocom, C. F. & Harris, S. W. 1966. Growth rates of Great Basin Canada geese. *Murrelet* 47: 33–37.

Zera, A. J. & Harshman, L. G. 2001. The physiology of life history trade-offs in animals. *Annual Review of Ecology and Systematics* 32: 95–126.

Zillich, U. & Black, J. M. 2002. Body mass and abdominal profile index in captive Hawaiian geese. *Wildfowl* 53: 67–77.

Index

Sub-headings in *italics* indicate tables and illustrations.